Deep

EARL BOEBERT
JAMES M. BLOSSOM

Deepwater Horizon

A SYSTEMS ANALYSIS OF THE
MACONDO DISASTER

Harvard University Press   Cambridge, Massachusetts, and London, England
2016

First printing

Library of Congress Cataloging-in-Publication Data

Names: Boebert, Earl, 1939– author. | Blossom, James M., 1950– author.
Title: Deepwater Horizon : a systems analysis of the Macondo disaster /
    Earl Boebert, James M. Blossom.
Description: Cambridge, Massachusetts : Harvard University Press, 2016. |
    Includes bibliographical references and index.
Identifiers: LCCN 2015051243 | ISBN 9780674545236 (hard cover : alk. paper)
Subjects: LCSH: BP Deepwater Horizon Explosion and Oil Spill, 2010. |
    Deepwater Horizon (Drilling rig) | Offshore oil well drilling—Accidents—
    Mexico, Gulf of. | Oil wells—Mexico, Gulf of—Blowouts. | Offshore oil well
    drilling—Accidents—Prevention. | Offshore oil well drilling—Safety
    measures. | Oil wells—Blowouts—Prevention.
Classification: LCC TN871.3.B64 2016 | DDC 363.11/9622338190916364—dc23
LC record available at http://lccn.loc.gov/2015051243

This work is respectfully dedicated to those who were lost:

Jason Anderson

Aaron Dale Burkeen

Donald Clark

Stephen Curtis

Gordon Jones

Roy Wyatt Kemp

Karl Dale Kleppinger Jr.

Blair Manuel

Dewey Revette

Shane Roshto

Adam Weise

And all others who bear the scars.

Contents

Foreword

This is an extraordinary book that digs deeply into the demise of the *Deepwater Horizon*. Readers are likely to be On Edge in every chapter. Although this book is a factual account, it reads somewhat like a novel in being such an unusually detailed, thorough, and authoritative analysis of a disaster. It also enumerates many realistic precautions, each of which could have helped prevent the Macondo disaster. In reality there was no one weakest link; instead there were many weak links, and attention to them could have avoided what happened.

The book is also unusual in the ways it explores the depth and breadth of the causal factors that can be identified throughout—involving many layers of corporate and operational personnel, and multiple factors relating to technology, management, standard practices that do not adequately cover contingencies, and much more. In this analysis of the Macondo case, these factors are clearly multidimensional, multifaceted, widely distributed, and crying out for the retrospective analysis that this book achieves.

Many lessons are here for everyone involved in the exploration and production of oil and gas. But much deeper, this book is an incisive parable for almost everyone involved in risky endeavors, even in completely different areas. It stresses the importance of planning for disasters, establishing detailed monitoring practices, carefully documenting instructions for seemingly routine operations, and even more important, carefully documenting changes in what must be done to anticipate and respond to possible effects— especially whenever the risks happen to be greatly increasing in real time. Even though it can be very difficult to realistically assess dynamic changes in risks in real time, it is absolutely essential.

The book's notion of the need to establish a pervasive and properly enforced "safety culture" is very timely in a world that emphasizes cost reductions and short-term optimization, to the deprecation of safety measures. Many organizations depending on life-critical systems might claim that they already have a safety culture, but it requires much more than lip service—it requires deep awareness of issues such as those considered here.

Furthermore, a similar observation also applies more generally to the need for a culture of predictable dependability and trustworthiness, whether or not human safety is a primary issue. Many computer-related endeavors require much greater reliability, resilience, security, privacy, and other mission-critical desiderata. In almost all disciplines, holistic thinking that encompasses concerns such as those considered in this book—and many more—is becoming a lost art. The need for ubiquitous risk awareness and risk avoidance (not just "risk management"), rigorous system practices, preventive maintenance, and many other factors is increasingly being widely ignored or given a much lower priority, typically in the quest for greater profits. As a consequence, the lessons of this book are enormously important in most technologically based enterprises, and are vitally compelling.

Peter G. Neumann

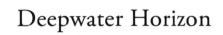

Deepwater Horizon

The *Deepwater Horizon* offshore oil rig and supply vessel *Damon B. Bankston* over the Tiber well in July 2009. *(Photo credit: iStock.com/Bradford Martin.)*

CHAPTER ONE

Introduction

THE EVENING OF APRIL 20, 2010, was a pleasant time to be out on the Gulf of Mexico. The seas were calm, the temperatures were in the seventies, and a light breeze was blowing out of the southwest. A quarter moon was low in a partly cloudy sky. Almost fifty miles out, the floating platform *Deepwater Horizon* hovered free above a mile of tubular riser that connected her to a well called Macondo. Beside her was the support vessel *Damon B. Bankston,* which had earlier taken on a large load of the heavy drilling fluid universally known as "mud."

About a mile away, the sport fishing boat *Ramblin' Wreck* was carrying three men on their first fishing trip of the year. One of the fishermen saw an unusual light in the distance. Through his binoculars he could make out a column of fire. Then a woman's voice came over the emergency channel of the VHF radio: "Mayday. Mayday. Mayday. This is the *Deepwater Horizon.* We are on fire. There has been an explosion. We are abandoning the rig."

The men aboard the *Ramblin' Wreck* headed toward the light, assembling along the way life jackets, first aid kits, floating pillows—anything that could be of use when something terrible has happened on the water.[1]

PRELUDE

It had been a busy night for the on-duty portion of the 126 men and women on board the *Deepwater Horizon,* perhaps the busiest ever. The *Horizon* was preparing to depart Macondo for her next well, and her crew members were eager to get there. They had completed the drilling phase eleven days before, and the drilling crew was now making the well safe for disconnect from the *Horizon* and for reconnect by another rig that would complete Macondo into a producing well. One of the *Horizon*'s cranes was laying out and stowing the numerous lengths of drill pipe and casing used in drilling, special crews were down below cleaning the tanks that had contained the drilling mud, and another crew was preparing a fitting to go on top of the well.

Adding to this activity was a visit by two executives from Transocean—the owner of the *Horizon*—and two from BP, which had contracted the rig and its crew from Transocean to drill Macondo. The executives were there to congratulate the crew on seven years without a lost-time accident. At 9:30 p.m. the executives were being entertained on the bridge as off-duty crew members slept, watched television, or caught up on their administrative work. The galley was open and the laundry was busy. A normal, if somewhat hectic, shift proceeded as so many had before.

More than three miles below, the bottom of Macondo was surrounded by hydrocarbon gas at pressures nearing 13,000 pounds per square inch (Figure 1.1). The previous night the drill crew had installed a cement barrier at the bottom of the well to seal off that gas. The operation had not gone smoothly. The crew gave the cement time to cure, and then around dinnertime on April 20 they conducted a well integrity test called a *negative test.* This, too, met with difficulties and produced unusual results, prompting the crew to repeat the test. The Transocean employees who were operating the equipment and the BP onboard overseer then agreed that the test had been successful, and moved to the next step: a scheduled activity called *displacement.*

Before displacement, the well had been full of the mud used while drilling it. This mud acted to equalize pressures at the bottom and relieve the newly constructed barrier of significant strain. As a second precaution, the *blowout preventer*—a massive flow-control device the drill crew used to prevent the uncontrolled release of hydrocarbons—was in the closed position.

The displacement plan entailed replacing some of the drilling mud in the well and riser with lighter seawater, in preparation for capping the top of

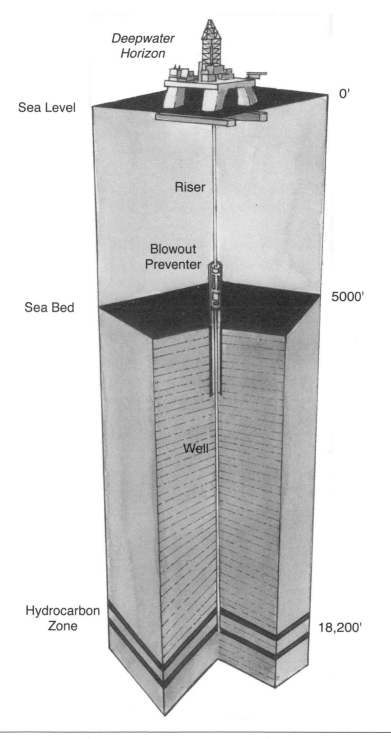

Figure 1.1. The *Deepwater Horizon* on station at the Macondo well.

the well and removing the riser. This required the blowout preventer to be open, so the mud could flow up and into storage tanks on the rig. However, so much mud was being removed that at some point there would not be enough left to equalize the pressure of the gas from the well, and the only thing holding it back would be the newly installed barrier. If the barrier held, there would be no problem. It failed.

BLOWOUT

No one will ever know the exact failure mechanism. What is certain is that the crew was unaware that massive amounts of gas from the well were heading toward them. By 9:30 p.m. the gas had risen above the blowout preventer, and nothing remained between it and the *Horizon*. As the gas rose and expanded, it pushed the mud and seawater in the riser into a fountain that flooded the *Horizon*'s deck and rained down on the *Bankston,* the support vessel.

A few moments later gas escaped into the calm night. The lack of wind allowed it to settle as a deadly, invisible blanket over the starboard and aft portions of the *Horizon*. Ignited by an unknown spark, it exploded into a pillar of flame that rose almost 500 feet into the air. Eleven members of the crew perished instantly. Power was lost, lights went out, and firefighting equipment was inoperable.

One hundred and fifteen people—some seriously injured—ran or were carried to lifeboats and life rafts. Some, including those who were delayed by an unsuccessful attempt to start emergency generators for firefighting, had to jump sixty or seventy feet into water covered with burning oil. The crew of the *Bankston* rescued every one of them.[2]

The seriously injured were evacuated by Coast Guard helicopter. The remaining members of the *Horizon*'s shattered community were forced to watch the platform burn for ten hours while the *Bankston* stayed on station as the incident command vessel, and then spent another fourteen hours under way to Port Fourchon, Louisiana.

On the morning of April 22, a day celebrated by environmentalists worldwide as the fortieth anniversary of the Earth Day movement,[3] the *Horizon* capsized and sank, breaking its riser and leaving Macondo open to the ocean.

Futile attempts to shut off the well by remote submarine and other means had started immediately after the rescue of the crew, and continued for weeks afterward. The well flowed unabated until crews finally managed to fit a temporary cap on the well in mid-July. In late September crews pumped drilling mud and cement through that cap and sealed Macondo permanently, forever entombing the evidence of what had failed.

The amount of oil that Macondo released into the Gulf of Mexico—and the corresponding environmental and monetary damages—may never be known. The human cost is known: eleven families lost a loved one, twenty-two children lost their fathers, survivors lost close friends, and many were scarred, mentally and physically, for life. The other costs began with the destruction of a half-billion-dollar drilling platform, extended to tens of billions of dollars in liabilities incurred by the corporations involved, and finished with widespread damage to the local environment and economy. Additional effort and money were spent on ten official and private investigations and the largest civil trial in American history. All in all it was—as one of the survivors wrote in his witness statement—"a very drastic event."[4]

THE EDGE

The transition of a complex system from a safe state to catastrophe is hard to predict, abrupt, and irreversible. A few minutes before 10 p.m. on the night of April 20, crew members of the *Horizon* were going about their business, on and in the giant machine that was both their workplace and their home. A few minutes after 10 p.m., they were trying to escape a nightmare of darkness, flame, pain, and terror. As one of the crew who struggled to save the rig later testified:

> Q. You've hit the function button for the Emergency Disconnect System, which should have—but did not—disconnect the *Horizon* from the riser so she could escape the burning well. Now that you've done that, you realize that the hydraulics may not be working, abandon ship has been called, you're leaving. What does your role become at that point?
> A. Survivor.[5]

Hunter S. Thompson likened the transition of a situation or system into disaster to what can occur when a motorcyclist seeking high-speed thrills rides along a twisting, dangerous highway:

> The Edge . . . There is no honest way to explain it because the only people who really know where it is are the ones who have gone over.[6]

And those who have gone over are dead, wounded, or damaged in other ways. Bikers and drillers both operate at greater or lesser distance from The Edge because both handle energy. Humans have needed energy to thrive for at least a quarter of a million years—since they first began to cook their food. The trick in using energy is to control its release: the difference between a smooth stop at a bikers' bar and a crash—or cozy gas heat and an explosion— is the speed at which the kinetic energy of a speeding motorcycle or the chemical energy of hydrocarbons dissipates.

The energy that drillers must control is in the form of hydrocarbon gas, which is poisonous, largely odorless, and highly explosive. It may be the prize itself, in the case of a natural gas well, or it may be a deadly partner of sought-after oil, as in the case of Macondo.

Hydrocarbon gas exists in porous strata of the earth—layers of sand-like material. It is guaranteed to be found at the depth known as *the pay zone*, where the sought-after hydrocarbons reside. It can also be found at more or less unanticipated strata along the way, always under great pressure. The energy in hydrocarbons comes in two forms: that generated from the heat and pressure of their location under the earth, and that released when they explode or burn. The *Horizon* fell victim to both: pressure pushed gas and oil up the riser and onto the rig, and a chance spark ignited it.

The most important safety requirement of a well construction project is maintaining *well control,* which involves holding hydrocarbons temporarily at bay while simultaneously constructing facilities to permit their controlled release and transport through refineries and other processing and ultimately to market.

The driller is the central safety figure on the rig. If the driller goes over The Edge, the rig can go with him: "Almost 20 years ago I stood on the rig floor of a semisubmersible with the Senior Offshore Supervisor I was working nights for . . . Pointing at the Driller on the brake,[7] he said 'In case you don't know, that is the most important person on this rig. He can sink this thing faster than anyone else on board.' "[8]

Drillers have traditionally avoided going over The Edge by exercising what is known in the U.S. military as *Fingerspitzengefül:* the "feeling in the tips of one's fingers." This is the ability to draw conclusions from incomplete and conflicting information and have the moral fiber to act. Drillers call it "listening to the well." They develop this intuition through a combination of innate ability and experience.[9] Such intuition is important not only for drillers on a rig, but also for managers on and off the rig who must approve decisions about what tasks to perform and how to perform them, and for engineers as well, lest they develop drilling techniques and rig designs that are infeasible or risky.

One critical factor differentiates drillers from bikers: bikers come close to The Edge only when they ride, whereas drill crews are near it whether they are boring through strata, finishing up, or waiting for cement to set. A motorcycle can be parked, but gas never sleeps.

Working in close proximity to potentially uncontrolled energy requires courage, and courage is prevalent on an offshore rig. Even the housekeeping staff must endure helicopter or boat rides in uncertain weather, and if the rig goes over The Edge, everyone on board goes with it.

But for those who are part of the well-control system, courage is not enough. They are in a situation resembling that facing pilots in the early days of commercial aviation. As Ernest K. Gann described, those who survived possessed more than simple courage: "A line pilot is *wary* all of the time, which is an entirely different matter. To be continuously aware, you must know what to be wary of, and this sustained attitude can only come from experience. Learning the nature and potentialities of the countless hazards is like walking near quicksand."[10]

A senior onboard supervisor assigned to the *Horizon* echoed this sentiment in a deposition during the trial related to the disaster: "You never trust the well, is my opinion, even though you successfully passed the negative test. I watch the well all the way through the unlatching of it."[11]

ANALYZING THE DISASTER

Individuals and institutions approach events of the nature and magnitude of the Macondo blowout with mindsets that can be characterized as either judicial or scholarly. The judicial mindset focuses on assigning or deflecting blame. The scholarly mindset seeks to understand a catastrophic event in order to prevent a recurrence. The judicial mindset concentrates on the accident that was, in support of assigning blame; the scholarly mindset considers both the accident that was and the accidents that might have been, seeking all factors with the potential to combine into a disaster.

In September 2010 BP released a report on its internal investigation, the first comprehensive report any organization had produced after the disaster. Perhaps not surprisingly, given BP's exposure to criminal and civil liability, its report manifests a judicial mindset.

The report identified eight barriers between the hydrocarbons in the well and the fire and spill, and grouped them into four critical factors:

1. "Well integrity was not established or failed." The barriers that failed were cement and associated mechanical barriers.
2. "Hydrocarbons entered the well undetected and drillers lost control of the well." Barriers that failed were pressure-integrity testing, well monitoring, and well-control response.
3. "Hydrocarbons ignited on the *Deepwater Horizon.*" Barriers that failed were surface containment of hydrocarbons and the fire and gas system.
4. "The blowout preventer did not seal the well." The barrier that failed was the blowout preventer emergency system.

The BP report further asserts that "if any of the critical factors had been eliminated, the outcome of the *Deepwater Horizon* events on April 20, 2010 could have been either prevented or reduced in severity."

The first critical factor was the responsibility of the cement contractor and the supplier of the mechanical barriers. The second factors were the responsibility of the Transocean crew. The third reflected the design of the rig. And the fourth was the responsibility of the supplier of the blowout preventer and the Transocean maintenance crew. BP itself is nowhere to be found, supporting the corporation's public stance that "this is not our accident but it is our responsibility to deal with it."[12]

The BP investigative team had adopted a variation of the 1930s domino model, in which a row of falling dominos represents a series of events leading to an accident. In the 1990s, James T. Reason developed a variation called the Swiss cheese model, wherein barriers fail in sequence and allow an adverse event to occur. Professor Nancy Leveson of MIT has this to say about such models: "The basic Domino Model is inadequate for complex systems . . . but the assumption that there is a single or *root cause* of an accident unfortunately persists as does the idea of dominos (or layers of Swiss Cheese) and chains of failures, each directly causing or leading to the next one in the chain. It also lives on in the emphasis on human error in identifying accident causes."[13]

The tendency to focus on a single cause is so common, and so misleading, that it has a name: "root cause seduction."[14] It is typically at the heart of efforts to assign blame or liability rather than prevent future accidents.

The shortcomings of a focus on root cause as a prevention tool can be illustrated by considering a hypothetical motorcycle rider who has learned the location of The Edge the hard way by charging ahead without a helmet, wearing dirty and scratched goggles, and on a motorcycle with poorly maintained brakes and a faulty rear tire. These factors put the rider close to The Edge before he or she even started out—a situation captured by the folk reference to "an accident looking for a place to happen." Let us further assume that accident investigators pointed to the rear tire as the "root cause."

A second rider, as casual about safety equipment and maintenance as the first, could read the accident report and be motivated to buy a new rear tire. Should that individual now ride without changing anything else, confident that he or she has eliminated the single "root cause"? Probably not, because the rider could be taken over The Edge by a different combination of bad brakes, bare head, and obscured vision.

This simple example may appear to exaggerate the way root cause seduction can divert attention and energy toward component-level "fixes" that are necessary but not sufficient to prevent a recurrence. The actual *Horizon* disaster shows a richer set of component-related risks spread across a variety of technologies and organizations; one expert in well construction enumerated thirty-five such "issues, factors, or questions" pertaining to Macondo, and official Coast Guard and Department of the Interior reports on the disaster list around fifty causal factors and multiple violations of international safety regulations.[15] Correcting just the subset of the risks associated with a single technology or corporation would still leave many others capable of combining into a trip over The Edge for those who ignore the possibility that they may exist.

Leveson further criticizes such "event-chain" models because their users rely on subjectivity to select both the events and the chaining conditions, and because the models underplay or ignore systemic factors. All these shortcomings—root cause seduction, subjectivity, and depreciation of systemic factors—are visible in both the BP report and other early reports that drew on its findings.

The scholarly mindset, in contrast, often adopts a *systems model*. A system is a collection of components developed more or less independently—plus the people who operate them. Two things make it a system: it has an intended purpose, and it exhibits what systems engineers call *emergent properties*. An emergent property is something that arises from the interaction between components rather than from the behavior of a single one. Emergent properties are things like safety, security, and reliability. They are typically important and hard to quantify.

Emergent properties suggest the possibility of *emergent events:* events that result from a combination of decisions, actions, and attributes of a system's components, rather than from a single act or from the failure of a single piece of equipment. The *Horizon* disaster was the very model of an emergent event.

The systems approach to staying away from The Edge involves detecting and forestalling emergent events before they arise. This, in turn, entails aggregating information from disparate sources, and focusing on interactions among the elements of the system as well as the state of each one.

Systems can be explicitly designed as such, in which case systems engineers will try to ensure that components work well together, and understand and control the system's emergent properties. In any system that entails hazards—because of either the materials it processes or the environment in which it operates—systems engineers will consider, in greater or lesser detail, how the system could fail and what the consequences of those failures could be.

Not only does such an effort reduce the chances that something bad will happen, but it provides an organized body of knowledge that the system's operators can apply to two problems. If they want to operate the system in a new environment, they can assess its shortcomings and make modifications. If something bad does occur, they can use existing knowledge to figure out what happened and why, and how to prevent a similar event from recurring.

Ad hoc systems, in contrast, "just happen": disparate groups assemble disparate components, and emergent properties arise more or less by accident. Even more important, the involvement of disparate groups leads to fragmented knowledge and little, if any, consideration of emergent properties before they emerge—an attitude that "it must be safe because nothing bad has happened yet." A common characteristic of ad hoc systems is the use of human beings as the integration mechanism. The system operates in a coordinated fashion only because of the individuals who are part of it: individuals who rely on tacit knowledge developed from apprenticeship and improvisation without the benefit of a unified systems concept.

The willful or accidental failure to analyze the emergent properties of an ad hoc system significantly increases the chance that something bad will happen when a system that is more or less safe in one environment is moved to another. In the case of the *Horizon* and similar rigs, several companies transferred practices, technologies, and equipment whose roots lie in land-based well construction to dynamically positioned ocean vessels without re-thinking fundamental challenges.

The fragmented and incomplete knowledge base underlying an ad hoc system also makes it hard to determine, after an accident, what actually occurred, creates uncertainty as to why, and provides less guidance on how to prevent a similar event. The *Horizon,* in its entirety, was manifestly an ad hoc system.

OUR APPROACH

We apply the scholarly mindset and systems approach to understanding how the *Horizon* disaster emerged. Leveson has proposed one of the most fully developed models for doing so. This approach sees an accident as the failure of a control system, which provides signals and other forms of information, and initiates action. In Leveson's words, "Safety is reformulated as a control problem rather than a reliability problem."[16]

Such a system is typically made up of technology, humans, and procedures and their intended and unintended interactions.

The Macondo blowout was a failure of a design and construction project—and of well control—that occurred before the permanent facilities were complete, like a building that collapses during construction. As such, it differed significantly from other well-known systems failures such as Bhopal and Three Mile Island, where a previously safe facility degraded through lack of maintenance or operator complacency. Because Macondo was in essence a failed engineering project, we must consider elements that are physically, temporally, and conceptually distant from the event itself, such as project management and organizational structure.

The fundamental structure of the offshore oil industry inhibits the aggregation of information, as it is scattered among technical specialties, corporations, and corporate entities. As a rule, the *operator,* such as BP, closely holds information on the geology penetrated by the well. The *contractor,* such as Transocean, is aware of the status of the equipment on the rig. *Oilfield service companies,* such as Halliburton, hold information on specialized technology such as that used to cement a well, and often keep elements of that proprietary. Organizational factors within the primary actor, the operator, may further complicate the construction of a systems view.

In the case of Macondo, a failure to aggregate information from disparate sources—and to take action based on that aggregation—allowed the emergence of an event that was wholly unanticipated at the time but inevitable in retrospect. A detailed study of how the event emerged provides not only insights into the causes of a failure to aggregate information, but also motivation for all those associated with offshore drilling to adopt a systems approach to safety.

Our Key Source: The Trial

When the scope of potential litigation arising from Macondo became clear, an agreement between plaintiffs and defendants consolidated 3,000 cases against various defendants and among 100,000 claimants into a federal non-jury trial known informally as MDL 2179.[17] Litigants in the trial had the right of discovery: to compel the release of records such as emails and corporate policy documents, and to call, depose, and cross-examine witnesses. This process yielded almost 4,000 trial exhibits and expert reports and 172 sworn depositions.

From February to May 2013, plaintiffs posted all material from the trial—including transcripts of testimony in open court—on the Internet. Of course, this material has been subject to selection bias, reflecting the Federal Rules of Evidence and the legal strategies of the parties. Despite those shortcomings, however, we analyzed the voluminous evidence to construct a more accurate and detailed, although still incomplete, account of the disaster than other analysts have provided.[18]

Notably absent from the accident is any form of event data recording, or "black box," common in fields such as commercial aviation. The only technical evidence on the Macondo blowout comes from a system informally known as Sperry Sun, which transmitted a subset of real-time drilling data to shore as an aid to drilling efficiency. The only physical evidence is the blowout preventer and some of its fittings to the well, retrieved from the bottom of the Gulf after flow from the well had been stopped. That means we must rely significantly on the accounts of survivors, many of whom chose to exercise their Fifth Amendment rights.

The lack of hard evidence on what transpired to push Macondo over The Edge makes us cautious about the conclusions we draw, particularly regarding what the people involved in a given control system *should* have realized. Forensic analysts operate in an environment that Professor Diane Vaughan of Columbia has dubbed "luxurious retrospective position."[19] We are looking backward, and know how the story ends. A signal that is clear to us may have been far from obvious at the time. No single individual may have been in a position to receive all the information or had the technical background to grasp its significance.

Still, we employ a systems view unrestricted by organizational or technological boundaries. Rather than attempting to reconstruct in detail what went on down in the well, we focus on the information that the system made available to people working in it, and how they responded to it. This analysis offers the most useful lessons to those on the front line of preventing a recurrence—those who must cope with difficult wells and the incomplete and often inconsistent signals provided by the technology used to drill them.

To maintain objectivity, we mostly refrain from making policy pronouncements, save the most basic: that all parties start viewing offshore drilling as a system rather than a disjointed set of technical and organizational elements—a system so complex that no one individual or entity truly understands it.

The trial evidence shows that the Macondo blowout emerged undetected through a combination of factors, including dangers inherent in drilling for hazardous hydrocarbons in the deep sea, BP's organizational philosophy, and inadequacies in the technology and management of the *Deepwater Horizon*. A final factor, unique to Macondo and one that complicated analysis, was an unrelenting pressure to push ahead in the last few days.

What Our Evidence Shows

A detailed examination of the events during the two weeks leading to the blowout reveals that those involved pursued activities at an ever-faster pace—on the final day almost as fast as physically and humanly possible. Every time it had an opportunity to pause, aggregate information, and assess how close to The Edge they were, the Macondo team declined to take that opportunity and instead chose to press on.

This is evidence of the psychological state that sociologists call "confirmation bias," pilots call "get-home-itis," and rocket scientists call "go fever":

the natural human tendency to be optimistic and interpret evidence in the most positive way. It has been shown to be the driving force for many a trip up to and over The Edge.[20]

Almost all the investigations of and commentary on the Macondo blowout have recognized the unusual and dangerous haste of those last days. However, these analysts have assumed that go fever either arose spontaneously or reflected simple greed on the part of BP: a desire to, as soon as possible, stop expending the million dollars a day Macondo cost them. Neither explanation survives close scrutiny.

It is true that oil companies are under constant cost pressure because they have no effective mechanism for passing costs onto customers. This may seem odd to the average consumer who is exposed to price fluctuations at the gas pump. These prices reflect market forces for the downstream segment of the business, which includes refining and retail sales.

The other major segment of the industry, *upstream,* includes exploration and production of crude oil. Upstream produces the bulk of oil companies' revenue, including 85 percent of BP's revenue in 2009.[21] In this arena, oil companies do not have the option that manufacturers do of justifying a more expensive product to potential purchasers by including extra features or exhibiting higher quality of manufacture. Commodities markets determine the price of each of roughly 200 grades of crude oil independent of which company produced it. It is therefore tempting to cite the simple desire to save on exploration costs to explain the disaster.

However, efforts to curb Macondo's cost do not explain the performance of BP and the *Horizon* before Macondo. Together they had drilled almost fifty wells without incident, including one that set a record for depth of water and depth of drilling. Any plausible explanation for the *Horizon* blowout must show why it occurred at Macondo but not at any of the previous wells. Our evidence shows that the explanation lay 200 miles to the southwest, far out in the Gulf of Mexico.[22]

Oil companies must find new oil to replace the oil they sell or they are in danger of going out of business. BP's strategy was to concentrate that effort on a few high-risk "supergiant" deposits. One was the *Horizon*'s next assignment: a monster discovery called Kaskida—a once-in-a-lifetime opportunity sixty times as large as Macondo.

BP was at risk of losing its lease to Kaskida if the company did not meet a regulatory deadline for drilling there, and the *Horizon* was the only rig BP had available for that task. BP executives have steadfastly denied that Kaskida was a factor in the blowout, and both official and unofficial investigations have accepted their denials. Despite that, we believe that such an extreme and widespread epidemic of go fever could arise only from the fear of losing a prize the size of Kaskida.

We hope our effort will help oil industry executives and engineers, federal and state officials, researchers, environmentalists, and other stakeholders separate the risks inherent in the very business of extracting hydrocarbons in deep water from those that some combination of political will and technology can mitigate—to ensure that decisions at every level minimize the chance of a recurrence. That contribution is critical because a wave of layoffs and retirements in the industry—the so-called Great Crew Change, now accelerated by the decline in oil prices—may force some individuals to assume responsibility for drilling wells in highly dangerous deep-sea situations before they have developed *Fingerspitzengefül*. We hope that group benefits from our work.

But the lessons of the Macondo blowout are not limited to the offshore drilling industry: they apply to any enterprise that operates in environments where catastrophic events can emerge from seemingly benign elements. We hope our narrative benefits future investigators and researchers into systems safety in such conditions, and provides insights into how the complex systems required to pursue such work can fail.

CHAPTER TWO

Controlling Macondo

THE MACONDO PROJECT was controlled by a hierarchy of systems, loosely associated with one federal agency. At the top was corporate BP, headquartered in London. At the center were two closely coupled systems: a BP organization in Houston officially known as Exploration and Production and unofficially as "Town," and the *Deepwater Horizon* itself, informally known as "the rig." Town—part of BP's Gulf of Mexico operation responsible for well construction projects—controlled the rig, and the rig controlled Macondo. (See Figure 2.1.)

Town submitted applications to construct wells and plans to proceed or modify these plans to the federal Minerals Management Service (MMS) the primary regulatory body overseeing well construction in the Gulf of Mexico.[1]

Town sent tasks and procedures to the rig by telephone and email several times a day. These were derived from Town's other activity: designing the well and assessing its geologic formations as the crew was drilling through them. People on the rig, in turn, carried out these tasks while using the technical facilities of the rig to continually maintain control of Macondo.

Two organizations had the ability to set limits on the behavior of Town: corporate BP and MMS. Neither did so to any great degree, which com-

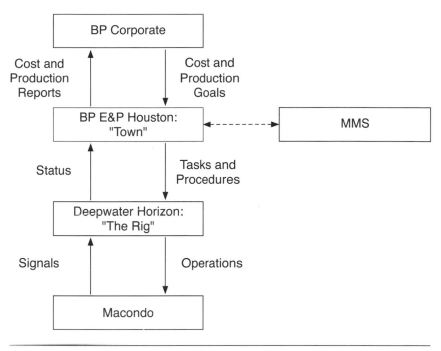

Figure 2.1. The hierarchy of systems that controlled Macondo.

bined with other factors to permit a looseness of discipline that oilfield workers refer to as the "cowboy attitude."[2]

BP CORPORATE

Of the many parameters by which we can measure corporations, two are significant in our control system model of safety: how a company organizes its technical assets, and how much initiative it allows lower-level entities. Corporations that must cope with frequent changes in technologies and markets, such as financial firms and consumer product companies, typically adopt a decentralized structure that distributes technical assets among operating entities that are granted a high degree of initiative. These entities are run, to the maximum possible degree, as if they were independent companies. They are measured on what they achieve and have few constraints on how they achieve it.

In contrast, oil companies—whose market structure and technologies are relatively stable, but which operate in an environment where catastrophic events are possible—typically concentrate technical assets in centers of power called "functional organizations." These subunits support safety in two ways: they maintain technical capability and corporate memory in specialized areas, and—if powerful enough—provide a pathway for transmitting bad news to upper management. Such news might include information that a well such as Macondo poses unacceptable risks.

Exxon Mobil provides an extreme example of this approach. In its soul-searching after the massive *Exxon Valdez* spill in the Gulf of Alaska in 1989, the company imposed the *Operations Integrity Management System* (OIMS), a set of rigidly enforced standards that apply to every step of every operation that the company undertakes.

Until 1990 BP organized itself around the "matrix model" prevalent in the industry, which similarly drew on strong functional organizations. Then everything changed.

A Fateful Reorganization

In 2010 BP's operation in the Gulf of Mexico reported to a corporate head-quarters in London that many people still called British Petroleum,[3] whose nominal lineage went back to the founding of the Anglo-Persian Oil Company in 1908. However, in reality BP became a twenty-first-century American company when it acquired Chicago-based Amoco in 1998 and Los Angeles-based ARCO in 2000.

In March 1990 the CEO who would later engineer those acquisitions had torn BP apart and rebuilt it in the model of an entrepreneurial financial firm. He divided the company into highly independent organizations eventually called *strategic performance units* (SPUs). Each SPU ran like an independent business, one that was gauged—and compensated its executives—on adherence to production goals and cost constraints dictated from corporate headquarters.[4]

The CEO's reorganization also eliminated the company's strong functional organizations in technical areas such as drilling, cementing, and other specialties, which held institutional memory in those safety-critical arenas. Under the new approach, BP outsourced most technical support for oil-drilling projects to subcontractors.

The company deployed its few remaining technical experts as internal consultants called *sector specialists*. Sector specialists had no authority over individual projects and no effective reporting chain to escalate concerns, and provided technical support only when project leaders requested it. For example, one sector specialist provided support for cementing throughout BP's entire Western Hemisphere operations—one of twenty-five responsibilities he was expected to fulfill.[5] A second sector specialist supported cementing for operations in the rest of the world.

As one business journalist with extensive experience in the oil business described it: "BP was no longer a technical company, but a commercial one. Its primary role was to decide which projects to back, and where possible to have others do the actual work."[6] The author of a 2002 Stanford case study and a 2004 book lauded this "disaggregated model," asserting that "real gains in performance can often be achieved by adopting designs that adhere to the basic logic underlying BP's disaggregated model. These include . . . giving strong incentives for unit performance . . . outsourcing . . . and, ultimately, the creation of a culture that is oriented to delivering performance."[7]

Neither work mentioned the dangers of hydrocarbons, the critical need for systems safety, or the importance of management of change, contextual reviews, and other tools of engineering management.

The organizational practices advocated by the Stanford author are appropriate for dynamic environments such as searching for new oil deposits, where the potential for gain is great and the consequences of failure are largely limited to the enterprise itself, in the form of failure to discover. As shown by prospects such as Kaskida, BP was very good at discovering new oil, and it is probably no coincidence that the CEO who embraced the Stanford practices and the successor who continued them were geologists by training.

The design and construction of offshore wells exhibit the opposite characteristics: they are tightly constrained by the nature of deep water, geologic formations, and hazardous hydrocarbons. Breakthrough performance is therefore unlikely, and Macondo showed that the consequences of failure can be a regional disaster. Oil companies that avoid such disasters do so by adopting the opposite model when constructing wells: strong and pervasive engineering discipline imposed from and enforced by the highest levels of the organization. Upper management at BP proved unable or unwilling to recognize the fundamental difference between finding oil and constructing wells to extract it; as a senior executive who left BP after the restructuring remarked later, "It never felt like an engineering company."[8]

When BP acquired Amoco and ARCO, it justified the wisdom of the acquisitions to the financial world by pursuing ruthless cost cutting. A longtime consultant to BP, called in to help consolidate the companies, said that he warned BP at the time about that strategy: "You're screwed. You just early-retired your memory. You early-retired the people who remember all those mistakes you ever made, and you've left all the bright young people without adequate mentors."[9]

BP's focus on fiscal performance was a success: it produced oil at the lowest cost per barrel of any of the "supermajor" oil companies.[10] As a result, it could pay shareholders high dividends—the traditional measure of performance for oil companies. That, in turn, made many people wealthy and earned a peerage for the CEO who was the architect of the company's new management structure. But those dividends carried what economists call *opportunity cost,* which is the lost benefit from activities and oversight that was not funded because the available money was spent elsewhere. In the case of BP, the opportunity cost of high dividend payouts was money *not* spent on maintaining facilities, keeping staff levels sufficient to provide resilience in the face of setback,[11] and engaging in the activities demanded by disciplined engineering management.

The Impact on Safety

The effects of these opportunity costs became clear as postmerger BP began sustaining accidents. In 2005 its refinery in Texas City, TX, exploded and burned, killing fifteen employees. The investigation by the U.S. Chemical Safety Board concluded that BP had "organizational and safety deficiencies in all levels of the organization."[12] In 2006 the company's corroded Alaska pipeline spilled 200,000 gallons of oil onto fragile tundra.[13]

According to the Center for Public Integrity, a watchdog group, the Occupational Safety and Health Administration cited BP 760 times from 2007 to 2010 for willful violations of worker safety at its refineries. Sunoco, with the next largest number of citations, received eight.[14] In all cases BP paid fines, settled lawsuits, expressed regret, promised to improve safety, and tinkered with its organization by adding committees and reporting requirements. What the company did not do was change its reliance on the cost of producing oil as its primary performance measure.

By 2009 the nimble entrepreneurial structure advocated by the Stanford case study had degenerated into a command approach as pathological as any that arose under the Soviets. Instead of a distant bureaucracy dictating production quotas with no concern for feasibility, BP corporate imposed equally arbitrary limits on costs to maintain profit targets established by headquarters.

Even when corporate dictated safety procedures, their implementation ran head-on into BP's fixation on cost, as noted by a consultant who coauthored a safety handbook for BP in the aftermath of Texas City: "They trained their refinery guys in the language of the book and then told them, 'It's up to you to implement this in each refinery.' . . . 'But, you have to do this within your budget, and by the way we're cutting your budget.' "[15]

For example, BP's June 2009 fiscal plan[16] directed the Gulf of Mexico SPU to cut its costs by 22 percent from 2007 levels by 2011.

The Impact on Macondo

Many investigators have searched for a smoking gun showing that BP knowingly put the *Horizon* and her crew at risk to save money. The most direct evidence came during a deposition on the *Horizon* disaster, when lawyers asked the chief operating officer for that SPU, "With respect to that cash cost reduction, were you involved in helping implement that in the Gulf of Mexico?"

He responded, "Not directly, no."[17]

Lawyers similarly asked an individual who had been abruptly dismissed at the end of 2009 from his position as the head of Exploration and Production: "Did anyone in BP's management ever tell you to choose cost over safety in 2009?"

After a six-second pause, he answered: "Not explicitly."

However, when pressed, the dismissed executive described his situation during the year before the explosion: "I can only—I can only state that during my time there in—there was—and I will use the term, in my personal opinion, incredible pressure on reducing costs."[18]

What we have found is implicit evidence of the pervasive effects of such cost cutting: understaffed subunits that gave "bright young people" responsibilities that exceeded their experience and deprived them of mentoring, and organizations that lacked the resilience to deal with absences of key personnel and as a result operated in an atmosphere of chronic crisis.

MINERALS MANAGEMENT SERVICE

Established in the early days of the Reagan administration, MMS had three charters: to generate revenue by leasing offshore federal land for drilling, to enforce safety regulations, and to protect the environment.

The agency was very good at the first charter, becoming for most of its history the second-largest source of federal revenue—only the Internal Revenue Service did better. The agency's performance on the other two charters, safety and environment, was considerably less impressive.

The National Commission on the BP *Deepwater Horizon* Oil Spill and Offshore Drilling, a blue-ribbon group cochaired by a U.S. senator and a former administrator of the Environmental Protection Agency, delivered this judgment of MMS to the president:

> Federal oversight of oil and gas activities in the Gulf of Mexico—almost the only area where substantial amounts of drilling were taking place—took a generally minimalist approach in the years leading up to the Macondo explosion. The national government failed to exercise the full scope of its power, grounded both in its role as owner of the natural resources to be developed and in its role as sovereign and responsible for ensuring the safety of drilling operations. Many aspects of national environmental law were ignored, resulting in less oversight than would have applied in other areas of the country. In addition, MMS lacked the resources and technical expertise, beginning with its leadership, to require rigorous standards of safety in the risky deepwater and had fallen behind other countries in its ability to move beyond a prescription and inspection system to one that would be based on more sophisticated risk analysis.
>
> In short, the safety risks had dramatically increased with the shift to the Gulf's deepwaters, but Presidents, members of Congress, and agency leadership had become preoccupied for decades with the enormous revenues generated by such drilling rather than focused on ensuring its safety. With the benefit of hindsight, the only question had become not whether an accident would happen, but when. On April 20, 2010, that question was answered.

The actions of MMS during the Macondo project consisted of quickly approving all the applications BP submitted, permitting a project plan that did not include any consideration of a blowout,[19] and accepting an outsourced oil spill response plan[20] that not only listed walruses as one of the species to be protected but also named as a technical resource a professor who had been dead for four years.[21]

The other impact of MMS on the Macondo project was both indirect and negative. BP's fear that MMS would revoke the lease to the Kaskida prospect—whether justified or not—spurred the company's fatal rush to get off Macondo.

CHAPTER THREE

From Rig to Well

DEEP-HOLE DRILLING EQUIPMENT originated in the Sichuan province of China more than a thousand years ago. Drillers pounded bamboo pipe to depths of 2,000 feet in search of salt in the form of brine. They erected a derrick to hold the pipe upright, and a winch, or *draw works,* to raise and lower it. Those three components—derrick, pipe, and winch—persist to this day.

Around 1900, drillers began using rotary drilling, which entailed rotating a long drill string of pipe with a bit on the end that ground through the earth. This setup required a mechanism to turn the pipe—either a rotary table at ground level or a top drive higher in the derrick.

Rotary drilling produced cuttings, which had to be removed from the hole. Some unknown engineer determined that drillers could flush these out by pumping mud down the hole. The result was the addition of a *circulating system.* The circulating system is the primary means by which rigs communicate with wells.

Rotary drilling permitted the digging of deeper holes through more fragile formations—which in turn required reinforcing the walls of an open hole with high-strength tubing, fixed in place by modified forms of Portland cement. In the 1920s an inventor and entrepreneur named Halliburton began the tradition in which a specialized service company furnished and installed that cement. Also in the 1920s drillers installed the first manually actuated *blowout preventers* (BOPs) to shut off blowouts, or "gushers," in the relatively shallow and low-pressure wells of the day. As wells became deeper and pressures higher, the role of the BOP changed but the name persisted. With the addition of the BOP, the basic elements of a drilling package were complete. Refinement and adaptations since then have enabled drillers to move this package offshore and into ever deeper water.

WHERE PUSH COMES TO SHOVE

Drill crews stay on the safe side of The Edge by practicing *well control*—the art and science of holding pressurized liquids such as brine and hydrocarbons at bay while drilling and constructing a well through them. That pressure, called *pore pressure* or *formation pressure*,[1] is produced by the weight of the layers of earth above the porous strata holding the liquids. Pore pressure can be significant—pressure at the bottom of Macondo reached almost 13,000 pounds per square inch (psi)—and when it is released by drilling into the strata, it will force the pressurized liquids upward toward the lower pressures on or near the surface.

The drilling community recognizes two forms of well control. *Primary well control* uses a column of drilling mud to generate enough pressure to resist the pore pressure at a given depth.[2] Drilling mud consists of a base fluid mixed with a heavy mineral such as barite. Varying the amount of mineral changes the density of the mud—and therefore the pressure it generates—to compensate for different pore pressures. Drillers use a variety of measurements to "listen to the well" and select an appropriate density of mud for the pore pressure they are encountering and the depth at which they are encountering it. BP and government regulators measured density in pounds mass per U.S. gallon (ppg), with mud ranging from 8.5 to a possible (but rarely used) 22 ppg. *Secondary well control* uses the adjustable barriers on the BOP to isolate the pressure from the rig.

When the "push" of the column of mud is sufficient to control the "shove" of the pore pressure, the well is said to be *overbalanced* or *static*. Conversely, if the push is not sufficient, the well is *underbalanced*. An underbalanced

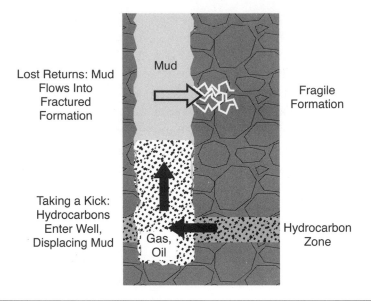

Lost Returns: Mud
Flows Into
Fractured
Formation

Mud

Fragile
Formation

Taking a Kick:
Hydrocarbons
Enter Well,
Displacing Mud

Gas,
Oil

Hydrocarbon
Zone

Figure 3.1. Well-control incidents.

state can allow pressurized hydrocarbons to enter the well from porous and permeable strata. If secondary well control does not stop that entry, the result could be a blowout.

When hydrocarbons enter a well during active drilling, this is known as *taking a kick*. We will use that term to signify an entry of hydrocarbons requiring immediate attention by the drill crew, while bearing in mind that the Macondo kick and subsequent blowout resulted from a failure of the barrier used to isolate the well from the formation. The latter is highly unusual and subtly different from a kick during drilling.

Primary well control is complicated by the tendency of formations to break down under the pressure of the mud, allowing the mud to escape from the well and into the formation. This phenomenon is called *lost returns*. Both kicks and lost returns are referred to as *well-control incidents*. The two classes of incident are illustrated in Figure 3.1.

Macondo was a particularly fragile well, and the drilling of it was plagued by lost returns. Some 16,000 barrels of expensive mud—an unusually large amount—disappeared into the formation,[3] causing the crew and drilling engineers to fixate on lost returns at the expense of other risks, such as a blowout.

When primary well control proves inadequate, as when a kick arises, drillers invoke secondary well control by exercising the facilities of the BOP.

THE CIRCULATING SYSTEM

To the average driller, the circulating system is one part of the equipment used to construct a well. To us, it is the control loop that connects the driller to the well, the central technology supporting well control, and the bottom layer of the systems model we presented in Chapter 2.

The circulating system on the *Horizon* was comprised of storage tanks (called *pits*), pumps, and sensors connected by piping. The pits—originally simple holes in the ground—are used to temporarily store drilling fluids. During active drilling, the circulating system pumps mud out of a pit, down into the hole, and then back up, where it is captured in the same or a different pit.

The *Horizon*'s circulating system was connected to a mile-long riser that dropped to the seabed and connected to the adjustable barriers of the BOP, as shown in the model in Figure 3.2.

The state of interest in this model is whether the well is static or taking a kick. During Macondo's last critical hours, the crew could not measure this state of interest directly, but instead had to infer it from signals provided by the most basic of sensors—all of which were sitting more than 18,000 feet above the likely point where hydrocarbons were beginning to enter the well. These sensors measured pressure, the volume of fluid in various pits, and the rate at which fluid was flowing out of the riser and into the pits selected by the driller. In this simplified model, sensor readings are displayed to the driller and transmitted ashore by means of the Sperry Sun telemetry system. The actual system was more complex and involved more individuals, as we describe in Technical Note 6 in the Appendix.

The driller on the *Horizon* had two command interfaces to the overall system: one associated with primary well control, and one associated with secondary well control. He issued some primary well-control commands electronically, such as starting and stopping pumps and adjusting their rate. These commands could prompt the system to send fluid down the riser, through the BOP, into the well, and back up again. Other primary well-control commands, such as selecting which pit the drilling mud should come from,

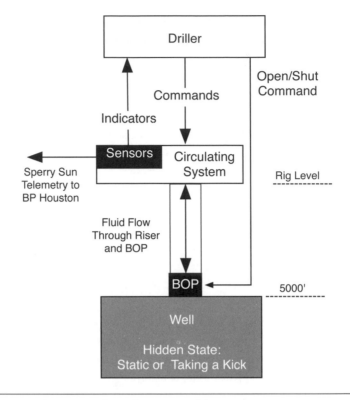

Figure 3.2. General structure of the circulating system.

required dispatching a *floor hand* to manually open and close valves. Secondary well-control commands were restricted to selecting a barrier in the BOP and commanding it to be open or closed—commands that the driller executed through a separate panel dedicated to that purpose.

The primary and secondary well-control facilities of the circulating system monitor and change the state of the well. For example, a driller facing lost returns could use the primary well-control facilities of pumps and pits to replace a given weight of mud with a lighter one, to reduce the amount of pressure on the well produced by the column of fluid. However, that approach would also reduce the amount of pressure available to resist the pore pressure at the bottom of the well, and could allow a kick. The driller would detect a kick by consulting displays showing measurements from sensors on the returning fluid. If he concludes that a kick is under way, his option is to use the secondary well-control facilities: closing one or more barriers of the BOP, thereby isolating, or *shutting in,* the well. At this point the crew has stepped back from The Edge and can take time to diagnose what caused the well to enter a dangerous state and how to bring it back to a static state.

LISTENING TO THE WELL

The most basic and reliable indication of a kick comes from the pit volume sensors. If the crew operates the circulating system as a closed loop—that is, the fluid pumped out of the pits is returned to the pits—the pits' volume should remain constant as drilling fluid circulates. That is, the amount of fluid coming back should equal the amount pumped in. If pit volume increases, then extra fluid is coming into the well, most likely from a hydrocarbon-bearing formation.

The gold standard of kick detection is the *flow check,* or the *flow test.* The principle is simple and robust: the driller stops the pumps for a period of time, and observes the return flow by means of a sensor or visual observation of the fluid coming out of the riser. After a short period, the return flow should stop. If it does not, then something is entering the well and pushing the fluid up the riser.

Because of the distance between events in the well and the sensors whose signals reflect them, the signals are often mixed with a greater or lesser amount of noise. Experienced drillers apply a kind of methodical wariness to this problem, in which they perform predictive calculations or observations before initiating an action, and watch to see if the signals match the prediction. For example, in a procedure sometimes called "fingerprinting the pumps," drillers record the characteristic way the flow for a given pump diminishes when the pump is stopped. If a flow does not drop off as quickly as the fingerprint suggests it should, that could signal a kick.[4]

THE ROLE OF THE BOP

The term "blowout preventer" was probably one of marketing genius when the devices were first invented and deployed on land in the 1920s. Like many marketing terms, though, it can mislead, because it gives the impression that the BOP is a dedicated safety device. This impression is reinforced by the fact that one of the adjustable barriers on the *Horizon*'s BOP was called the *blind shear ram* and was designed to cut through drill pipe and mechanically seal the well.

The limitations of the technology compared with the pressures that drillers can encounter in a modern deep-sea well make applying the term "blowout preventer" to the adjustable barriers that protected the *Horizon* from Macondo's hydrocarbons as misleading as calling the brakes on our biker's motorcycle a "collision preventer." Brakes can indeed be used to prevent collisions, so the term is not patently false. But they are also used in many other situations, and eventually require maintenance. Also, they prevent collisions only when used in time. And they are seldom tested in full emergency conditions, so trust in them is often based more on faith than experimental rigor.

The *Horizon*'s BOP shared most of these characteristics, including prevention as one of many functions, the need for continual maintenance, the requirement for timely initiation, and the lack of testing under full emergency conditions. The one capability of the BOP that motorcycle brakes usually do not have was to automatically sequence emergency actions—a capability that ultimately failed.[5]

The fact that drillers view the BOP primarily as a well-control device and only secondarily, if that, as an emergency mechanism was pointed out shortly after the blowout by the president of a drilling company, in reaction to assertions made by the CEO of BP: "[The CEO] and BP have taken the position that this tragedy is all about a fail-safe blow-out preventer (BOP) failing, but in reality the BOP is really the backup system, and yes we expect that it will work. However, all of the industry practice and construction systems are aimed at ensuring that one never has to use that device. Thus the industry has for decades relied on a dense mud system to keep the hydrocarbons in the reservoir and everything that is done to maintain wellbore integrity is tested, and where a wellbore integrity test fails, remedial action is taken."[6]

The use of BOPs in offshore drilling reinforces this observation. Application of the blind shear ram in a well control or other emergency is exceedingly rare. A senior supervisor on the *Horizon* told government investigators that in the sixty to seventy-five wells he had drilled, he had never seen a blind shear ram activated in a well-control situation.[7] This fact, along with our study of drillers' conversations on Internet sites, supports the notion that drillers are careful to avoid situations where they must use the BOP's blind shear ram to try to stop a blowout—just as prudent motorcyclists take pains to avoid situations where they have to slam on their brakes at high speed to prevent a collision.

This is a different attitude from that taken by many investigators and media outside the world of well construction, who have placed great emphasis on the shearing capabilities of BOPs. Although making these facilities more robust would be beneficial, a disproportionate focus on them risks the pathology of mistaking a necessary improvement in a component for a sufficient systems solution—a classic manifestation of root cause seduction.

BOP Functions

The *Horizon*'s BOP had an array of adjustable barriers that were controlled from a panel in the drill shack. Of these, only two played a significant role in the events leading up to the blowout: the upper and lower *annular preventers*[8] that were the topmost in the array. These devices operated like an adjustable elastomeric "donut" that could shut in the well by squeezing around a drill pipe, or on itself if no pipe was there. In all the significant actions involving the annular preventers on the *Horizon,* they were closing on drill pipe. Also, for reasons that are explained in Technical Note 2 in the Appendix, the vast bulk of secondary well-control actions on the *Horizon* involved just the upper annular preventer. The physical arrangement of that device is shown in simplified form in Figure 3.3.

The drawing omits the complex mechanism of valves and pistons that squeeze the "donut" and force it to expand and fill the annulus. In the open position, fluid such as drilling mud can flow from the well through the BOP annulus and into the riser. When closed, as shown in the right-hand drawing, the flow is blocked and the well is "shut in" to the degree that the "donut" can resist any pressures generated within the well.

In subsequent diagrams depicting operation of the BOP we represent this mechanism as black rectangles that either allow or block the path between the riser and the well, as shown in Figure 3.4.

BOP Fluid Paths

Our simplified BOP diagrams will typically show one or both of two paths by which fluids such as mud can be pumped into the well. The first of these is drill pipe, which runs down the middle of the riser and the BOP. The second is an auxiliary line called the kill line, which runs down the outside of the

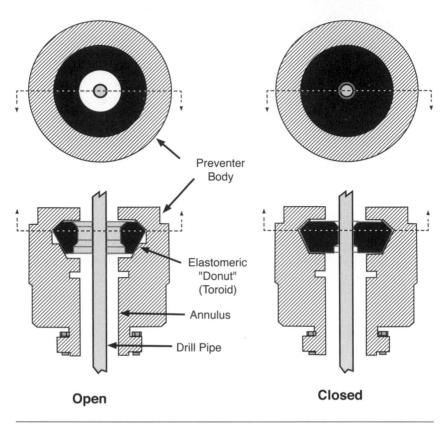

Figure 3.3. Operation of the annular preventer.

riser and enters the BOP just above the point where it is attached to the top of the well. The kill line is normally used in the recovery from a kick, as a means of pumping heavier mud into a well whose annular preventer has been closed. In the period we are concerned with, the crew used the kill line as an alternative path for testing the integrity of the well after the final cement job. During this period both the drill pipe and the kill lines had pumps lined up to them, and both had pressure sensors.

Both lines had valves on the upper ends to permit control of pressures transmitted through them from the well. In addition, the kill line had a remotely operated valve at the point where it entered the BOP. This valve was hydraulically controlled, and equipped with a spring that would close it if it were in the open position and hydraulic pressure was lost. A shortcoming of the BOP control system was that there was no means to signal the crew that

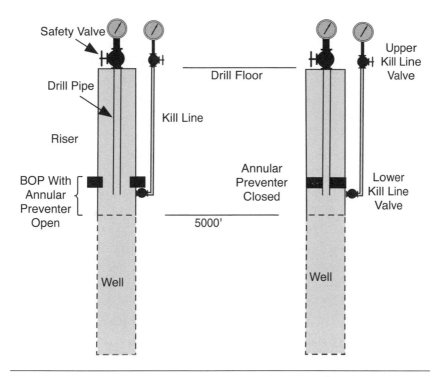

Figure 3.4. Symbolic diagram of the BOP and associated equipment.

the valve had automatically closed, a shortcoming of potential importance when it came to testing the integrity of the cement work at the bottom of the well—cement work that was, at one point, all that was between the hydrocarbons and the *Horizon*.

WORSE THINGS HAPPEN AT SEA

The consequences of a loss of well control are always severe, and are even more so when constructing wells offshore, a fact made clear by considering how an offshore system differs from the land-based systems from which the technology evolved.

Revolutionary changes in drilling technology are exceedingly rare. There is nothing in well construction to compare with the transition from vacuum tubes to solid-state circuits in electronics, or from piston to jet power in aviation. Rather, the industry takes what it already owns and understands and adapts it to new conditions on a component-by-component basis. Such an approach rests on the largely unstated assumption that the new environment will be only incrementally different from the old, so an incremental change to devices that worked in the old will suffice. The evolutionary nature of offshore technology is the result of the extreme technical conservatism of the oil industry—a conservatism driven by the capital-intensive nature of the business, the volatility of oil prices, and the hazardous nature of the activities.

In the case of deepwater work, whether this assumption holds up is by no means certain. The special problems posed by offshore work can be illustrated by considering how drilling crews in onshore and offshore environments maintain the BOP, and their options if they do not shut in a well soon enough and a kick turns into a blowout.

With land-based rigs, the BOP rests on the ground at the top of the well and is easily accessible for inspection, testing, and maintenance. Hydraulic leaks, marginal electrical connections, and unusual reactions to commands are immediately visible to the crew. Control and power cables are likewise out in the open. The accessibility of a land-based BOP not only improves the system's reliability by making potential problems easier to detect, but also removes inhibitions to fixing them, because the amount of time drilling must stop to enable the crew to repair the BOP is not much longer than the time they spend preparing for and performing the work on the device.

A blowout on a land-based rig threatens only the drill crew in the immediate vicinity of the well, and many blowouts occur with no casualties at all, because crew members can simply run a safe distance from the rig once they see drilling fluid rising from the well. And the sleeping quarters, galleys, and recreational facilities for the crew and support staff of a land-based rig are typically some distance from the well, out of harm's way.

After a well on land blows out, a specialist team can "kill" it with methods such as using explosives to blow out the flame, and then cap the well with a heavy fixture. This process, while spectacular and dangerous, is more or less routine for such organizations. When Saddam Hussein ordered the deliberate blowing out of more than 600 wells in Kuwait in 1991, specialist teams extinguished and capped all the wells in around eight months.

To reduce the environmental effects of a blown-out well on land, crews often deliberately ignite, or "flare off," the hydrocarbons it releases. Crews can also use earth-moving equipment to create berms or levees to contain any escaped oil.

Constructing wells in shallow offshore water from fixed platforms or anchored barges does not pose challenges that are significantly different from those entailed in drilling and operating land-based rigs. Even though the BOP in such offshore rigs may be underwater, raising it for inspection, testing, or maintenance is not very time-consuming, as cable runs are still rather short. If crews lose control of a kick and a blowout does occur, crew members can escape by jumping into the water or taking to lifeboats. And the typically short distance to shore enables crews to live there, so the only individuals at risk during a blowout are members of the relatively small crew on duty. Specialist teams can often—although not always—kill blown-out wells in shallow water on the platform or barge.

Things change significantly when well construction occurs tens of miles from shore and under thousands of feet of water. The *Horizon*'s BOP was separated from the rig by a mile of water, cables, and riser, and subject to external water pressure of about 2,200 psi. The depth at which the BOP sat meant that any significant repair could require several days of downtime on a rig that cost a million dollars a day to operate. The riser also held almost 1,800 barrels 75,000 gallons—of liquid, which delayed the transmission of information from the circulating system to the surface, such as the amount of hazardous hydrocarbon gas trapped in drilling mud.[9] And that riser effectively formed a reservoir that could store gas above the BOP during a blowout—gas that the crew had to deal with even if they closed the BOP immediately after detecting gas at the surface. When Macondo blew out, no organization or technology was capable of rapidly killing the well, so Macondo flowed for weeks.[10]

Finally, a rig such as the *Horizon* integrates accommodations for up to 148 crew members and service contractor personnel with drilling equipment in a single floating unit. The distance between the entry to the well and the farthest point in the accommodation spaces was about 150 feet, separated by fire-resistant but not blast-resistant bulkheads. The exploding gas destroyed those spaces, injuring seventeen people, several seriously.[11]

These challenges entailed in controlling deep-sea wells show how moving elements piecemeal to a new environment may result in an ad hoc system whose response to certain events may not become known until they occur. At that point it may be too late to prevent catastrophe—as was the case with the *Horizon* at Macondo.

CHAPTER FOUR

Weighing Macondo's Risks

ON MARCH 19, 2008, MMS sold the rights to drill for oil in the seabed known as the *Gulf Region of the Outer Continental Shelf,* netting the federal government a little over $3 billion. The boundary of this region stretches in a gentle southerly arc from the southernmost tip of Texas to just past the tip of Florida.

BP spent $34 million to lease a 5,700-acre block called Mississippi Canyon 252, or MC252. This block lay at a depth of just over 5,000 feet—48 miles from the nearest shoreline and 113 miles from the nearest usable port.[1]

BP had first surveyed this area in 1998, and surveyed it again in 2003. The company's geologists estimated that more than 60 million barrels of hydrocarbons lay some 18,000 feet below sea level.[2] Although this represented more than one-third of BP's yearly goal for discoveries in the Gulf of Mexico,[3] a BP geologist characterized the prospect as "small" and "not what we would consider a super giant" like Kaskida.[4]

BP originally named the first well in the block, officially MC252 #1, Epidote. In late 2008 BP held a charity contest for the right to rename the well, won by a Colombian-American group that chose the name Macondo, which was the name of a doomed village in the novel *One Hundred Years of Solitude* by Gabriel García Márquez.[5]

Most wells in the Gulf are categorized as *high-pressure, high-temperature* (HPHT). Macondo narrowly qualified as an HPHT well, with pressures at the bottom of the hole of about 14,000 psi, and temperatures of around 243 degrees F. HPHT wells in the Gulf typically combine high pressure with fragile strata, which complicates well control, and their fragility is difficult to measure.[6]

An expert witness for the plaintiffs during the trial described an area in the Mississippi Canyon called the *Golden Zone.* Subsurface temperatures in that zone range from 140 to 248 degrees F, which some geologists believe indicates hydrocarbon-rich formations—formations whose difficulty he believed the BP geologists should have foreseen.

"Drilling the Macondo 252 #1 well was rich in geological challenges and certainly not just an engineering nightmare. Drilling into the Golden Zone at this depth, temperature, and location should have been predictably challenging and required appropriate and thorough pre-drill modeling of pore pressure evolution, and great caution. Anything less inevitably would contribute significantly to the blowout disaster as seen on April 20, 2010."[7]

In fact, BP geologists seriously underestimated the fragility of the formation through which the crew would drill—one of several lapses in planning the well.[8]

THREE PURPOSES FOR ONE WELL

Exploiting a new prospect such as MC252 requires three distinct activities. The first is drilling an *exploratory well* to verify the existence of hydrocarbons and take initial measurements of the strata. Oil companies must pay special attention when drilling exploratory wells, assessing the largely unknown strata they will encounter on the way down. After verifying or disproving the existence of hydrocarbons and assessing the practicality of further development, companies often plug and abandon such wells.

If the exploratory well strikes oil, the next stage is an *appraisal:* assessing the volume and nature of the pay zone. Drillers do this by taking measurements from the exploratory well or by abandoning it and drilling a separate appraisal well.

The final activity is *production*—the long-term extraction of hydrocarbons. Production requires installing a *production casing,* a pipe inside the well that connects to the pay zone and permits the controlled flow of hydrocarbons into a production system, such as a pipeline to a refinery or a facility for tanker ships. As with the appraisal stage, oil companies can add production facilities to the existing well or drill another one.

The allocation of these three activities across one, two, or three wells depends on factors such as budget, rig scheduling, and the geology of the prospect. In the difficult geology of the Gulf, the conservative approach of drilling two or three separate wells is common.

Instead, from the start of drilling in January 2009, BP internally and informally characterized Macondo as a "keeper"—a production well.[9] In a report to his superior a BP engineering manager observed, "This well is very similar to Isabella [*sic*] which we installed as a keeper. There is a good argument for doing this one the same way."

Isabela was a nearby well of the same depth planned for Macondo. Despite this superficial analysis, the superior accepted the manager's argument, as shown by an email string from January 16, 2009, that examines the impact on the budget of including production-oriented steps in the initial project.[10]

When discussing Macondo with the National Commission, BP executives called it "an exploratory well with a production tail."[11] They meant that the exploratory well included drilling and appraisal, and that the production tail entailed installing production casing in the well. No one knew it in 2009, but the offhand decision to include the production tail would combine with the fragility of the formation and the pressures of Kaskida to take the Macondo project and the *Horizon* over The Edge.

WELLS OF INFLUENCE

Oil companies pay a *day rate* to lease drilling rigs from contractors such as Transocean, whether they keep the rig busy or not. *Horizon*'s day rate was roughly $1 million, including the cost of salaries and material. Because of that cost, oil companies are under pressure to ensure that every rig is fully used, and to monitor nonproductive time carefully. What's more, only a fixed number of rigs are available at any given time, and not all are suitable for constructing all wells, so assigning rigs to wells is tricky, especially if a well proves more difficult than anticipated, or weather intervenes.

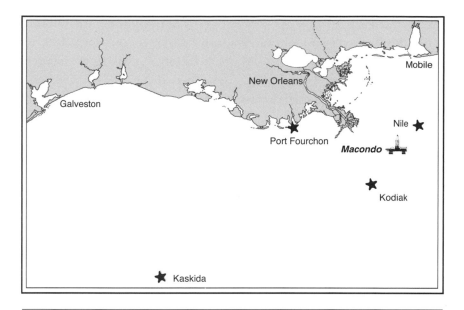

Figure 4.1. The Gulf of Mexico.

Three wells besides Macondo influenced the fate of the *Horizon:* a well called Kodiak, her previous assignment; and two upcoming projects: an appraisal well at Kaskida, and a "fill-in" job at a depleted well called Nile. The map in Figure 4.1 shows the locations of these wells.

The next scheduled activity for the Kaskida prospect was drilling an appraisal well before the regulatory deadline of May 16, 2010. This deadline arose from a requirement that operators begin work on a lease within a certain time after it was awarded, a requirement imposed to ensure that companies did not hoard or stockpile leases and thereby delay the payment of royalties to the government. If BP did not begin work at Kaskida by that date and MMS chose to enforce its own rule, the company could lose its lease on this mammoth discovery—a remote prospect in June 2009, but a pressing concern near the end of April 2010.

BP also faced a second regulatory deadline. The well called Nile had run out of oil in July 2009 and had to be plugged and abandoned before July 2010, under a rule that such wells had to be rendered safe within one year of the end of production. The risk there was not the loss of a lease but a regulatory black mark, although, as with Kaskida, the deadline appeared to be comfortably in the future in 2009.[12]

For reasons that none of the investigators of the blowout ever probed, when the *Horizon*'s schedule was squeezed for time in mid-April 2010, BP Town did not cut short work on Macondo by forgoing the production tail, or dropping the Nile work. Instead, Town chose to have the *Horizon* finish the production tail, proceed in the opposite direction from Kaskida to work on Nile, and then make the long trip to Kaskida. The schedule given in the *Horizon*'s rig planning spreadsheet[13] had the *Horizon* leaving Macondo on April 26. The history logged in that spreadsheet for the time to install and remove risers and BOPs and the transit time between wells shows that they would have only a handful of days to perform the work on Nile and accommodate any delays there, en route, or at Kaskida. This only worsened the pressure to finish Macondo.

BEYOND THE BEST

Part of the problem was BP's lax approach to engineering discipline. The company pursued the Macondo project according to a management process it called "Beyond the Best." A plaintiff expert witness and former BP consultant called this process "a primitive risk assessment and management system that could not pass muster in terms of requirements today for a process safety system that would enable management of a very hazardous system such as Macondo."[14]

BP attorneys presented this process, which consisted of stages, at MDL 2179. The stages and their contents were:

Appraise: Geological assessment and Risk assessment
Select: Analyze well design options, Approve well design, Update project risk register.
Define: Detailed design of the well, Independent peer review by BP technical experts, Update risk register.
Execute: Well plan is executed with adjustments made as needed based on conditions during drilling.
Review: Capture and disseminate lessons learned to the rest of the business unit.[15]

Each stage was to end with a "stage gate," depicted as a traffic light, with options to stop, proceed, or rework. The gate was to be controlled by a "gatekeeper" with authority to select among those options. No evidence reveals the gatekeepers for the Macondo project.

TWO TOOLS FOR ASSESSING RISK

The "risk register" cited in the BP's Beyond the Best process is one of two key tools that managers commonly use throughout risky engineering projects to estimate the distance to The Edge. The principle underlying risk assessment is simple: risk is the likelihood of an adverse event combined with the consequences of the event. The register is a table structured like that shown in Figure 4.2, usually in the form of a spreadsheet. Each box contains events of concern, ranked by their combination of likelihood and consequence. The darker the shading, the greater the risk.

The matrix in our example is typical of risk assessment in general, in that it is qualitative, a kind of structured *Fingerspitzengefül* that captures tacit knowledge, intuition, and experience. As such, it is a tool for sharpening thinking and avoiding oversights—not an oracle that will scout The Edge on its own. Its effectiveness as a decision support tool depends on how comprehensive it is—how many disparate signals and pieces of information it aggregates. It is deficient in that it does not capture the interrelationships between events, such as when an event that presents a moderate risk on its own may, if it occurs, raise the likelihood of a catastrophic risk—as when a splash of mud on a biker's goggles may be a minor risk by itself but more significant if it distracts the rider from seeing a "bridge out" sign. Despite these and other shortcomings, a risk register is a commonly used tool in a variety of environments because it is the best tool available.[16]

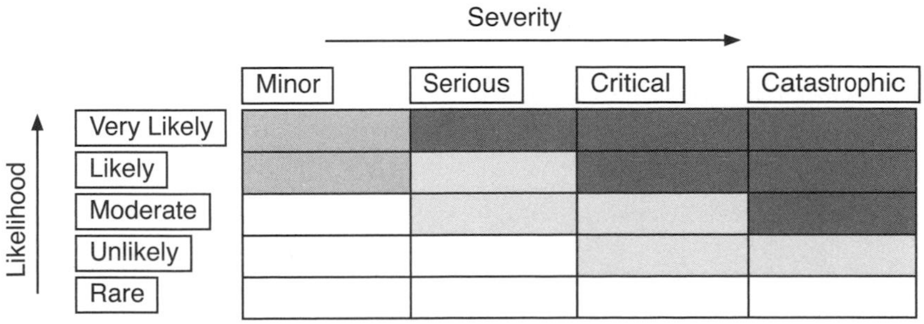

Figure 4.2. Format of a risk register.

The second tool is *management of change* (MOC), a process designed to prevent risks from slipping into a project as a result of inadequately scrutinized changes to procedures, designs, or materials. Like the risk register, it is based on aggregation of information, sober consideration of possible outcomes, and approval by management.

BP's Beyond the Best process did not mandate updating an oil well project's risk register after it passed the Execute Gate, and did not mention management of change at all. BP Town was aware of the need to assess risk and did so on occasion, but seems to have regarded any formal process as a bureaucratic hurdle to be overcome rather than an exercise that provided meaningful insights. And some Town employees gave the appearance of seeking ways to avoid the effort and delays associated with this process.[17] BP Town also relegated the mechanics of both the Beyond the Best process and management of change to an outside contractor, suggesting that it viewed both as externally imposed administrative burdens rather than an integral part of engineering management culture.[18]

These attitudes contrast sharply with Exxon Mobil's Operations Integrity Management System, which includes risk assessment and management of change as two of eleven elements. In its public literature on OIMS, the company cites a well called Blackbeard West as a case study. Exxon Mobil plugged and abandoned that well in 2006, writing off $187 million when unexpectedly high pressures in the well made it too risky to proceed.[19]

The BP legal team took pains to claim a similar management structure during MDL 2179, taking testimony from the man who led the Gulf of Mexico business unit from 2007 to 2009 and eliciting descriptions of BP's similar document called the Operations Management System and other process and systems safety directives. BP town had not fully implemented this document at the time of the blowout. And what the trial did not show—and what we did not find in any of the evidence—is any influence this or other safety directives had on the behavior of the individuals in Town.

The Macondo Risk Register

The Macondo risk register, which took the form of a spreadsheet,[20] was at the center of Town's "primitive approach to risk assessment and management." Macondo's risk register was particularly deficient as a tool for judging distance to The Edge. It enumerates a wide range of project risks but does not mention risks such as "blowout," "explosion," "injury," "death," and "oil spill."

Of twenty-three risks listed in the register, two concern risks to Macondo's long-term viability as a production well, five pertain to schedule, and the remaining sixteen focus on financial impact. The latest entry in the register is dated June 17, 2009, suggesting that it fell into disuse once the project had passed the Execute Gate.[21]

The register includes several risks associated with the fragile formation at Macondo. The primary step Town devised to mitigate those risks was to use foamed, lightweight cement to attach the bottom of the production casing to the formation, to reduce stress on the formation at that part of the well generated by pumping heavy cement slurry between it and production casing. The register records concern about possible failure of the cement, but in terms of cost—not that failure would trigger a blowout, as actually happened. The register also notes concern about the BOP on whatever rig is assigned to the well, but focuses on delay stemming from the need to repair the BOP, not that it could—as actually happened—fail to shut in the well during an emergency.

The sole entry on maintaining control of the well—The Edge for a drill crew—reads as follows: "Potential well control problem: risk of losing the wellbore in an uncontrolled situation."

The risk of losing a wellbore—that is, that an open hole will collapse in on itself if drilling pressures exceed the fracture pressure of the formation—is listed as a project risk rather than a safety risk. The matrix gives that risk an overall rating of "moderate," and includes a note that it is being addressed: "Casing program design to mitigate issues."[22]

The register mentions nothing more about losing control of the well. Ample evidence exists that the control of hazards such as a blowout must be considered explicitly and continuously from the very beginning of a project such as Macondo.[23] Instead, BP Town proceeded as if the possibility of a blowout did not exist.

The Macondo Peer Review

Although it did not specify it as such, in June 2009 Town conducted a drilling peer review that appears to have served as the Execute Gate for the project. In this activity, Town assembles a review team composed of experts from BP's Gulf of Mexico SPU to examine project risks, well design, and schedule and cost estimates.[24]

The Macondo peer review did not include well control among its primary drilling risks, although that concern appears in a condensed risk matrix. As with the risk register—and as would be the case throughout the well's exploratory and production tail phases—the review focused on lost returns as a risk to project costs or collapse of the wellbore, not explicitly to maintaining control of the well.

The plan the team reviewed covered both the exploratory phase and the part of the production tail that involved cementing the bottom of a single "long-string" production casing that was to run the full length of the well. The plan followed the risk register in specifying the use of foamed cement to hold the production casing in place at the bottom of the well, and in calling for drillers to slowly circulate that cement into position. The review presented the last two decisions as steps to offset the risk of fracturing the formation.

The review described the choice of where to put the Macondo well as driven by estimates of the location of *shallow gas*—pressurized strata near the surface that are difficult to control and a frequent cause of blowouts.[25] This consideration had led BP Town to put the well in a location unusual for the Gulf, in that its pay zone did not lie beneath a salt dome. That meant the strata through which the crew would drill were more geologically active.[26] Although BP had less experience drilling through such formations,[27] neither the risk register nor the review treated this location as posing a special risk. In fact, the review team noted that the absence of a salt dome was "conducive to a fast track schedule."[28]

Anticipating a Difficult Well

The plan's schedule and cost estimates were based on actual figures from twenty-seven wells drilled outside of salt domes in the Gulf's Mississippi Canyon. BP Town had calculated the mean number of days per 10,000 feet of depth drilled, and added 10 percent for nonproductive time, plus eighteen days of delay for hurricanes. The result was an estimate that Macondo would require ninety-eight days to drill and cost $120.6 million.

The plan specified that Transocean's *Deepwater Marianas* rig would begin drilling the well in August 2009. The *Marianas* was an old piece of equipment, having been launched in 1978 as the *MV Tharos,* a massive firefighting and rescue vessel designed by the legendary Red Adair, dean of the oil well firefighting community until his death in 2004. The *Tharos* had responded to the explosion and fire aboard the Piper Alpha platform in 1998 with mixed results, and the industry had dropped the idea of such a vessel. After several changes of owners and names, the *Tharos* was rebuilt as an anchored drilling rig in 1998 under the name *Deepwater Marianas.* (It is now known as the *Transocean Marianas.*)

Transocean rigs tended to specialize in constructing either exploratory/appraisal wells or production wells, but not both. The *Marianas* had never drilled an exploratory well, and BP had to add measurement-while-drilling equipment to support drilling in a location where no one had ever drilled before.

The *Horizon,* on the other hand, specialized in exploratory wells, and had installed production casings and fittings only four times in the previous twenty wells it had drilled.[29] And none of the BP supervisors on board during the rig's last days had supervised those operations. The peer review did not consider rig specialization and its effect on project and safety risks, and there is no evidence that it influenced the plan for drilling Macondo in any way.

Although the drilling schedule relied on the median time for nearby wells, the risk discussion in the peer review clearly indicated that Macondo was likely to be a difficult well. Following BP's Beyond the Best practice of setting performance targets slightly faster than the fastest historical times, the target schedule was fifty-four days. Whether those who had to do the actual drilling considered these targets realistic is unclear. The senior drilling engineer assigned to the project commented: "Pmean used to be the standard, but that is getting cloudy now that the focus is on performance . . . yeah, like the actual AFE number affects performance . . . give me strength."[30]

This comment suggests that those who prepared the drilling plan did not take it completely seriously, and that it would not provide a solid base for decision making if surprises occurred.

The plan did not subject the "exploratory well with production tail" approach to any scrutiny, and thus implied that successful exploratory drilling always includes a "production tail." But some formations are strong enough to allow crews to drill, plug, and abandon a well, but not strong enough to withstand the stresses of cementing production casing to the bottom of the well. The Macondo prospect showed signs of being one of them.

Despite all this, the outside reviewers delivered a positive assessment of the plan the very next day.[31] The reviewers praised the use of the long-string production casing, and observed that the pause for hurricane season would allow the team to "develop optimized work instruction and logistics" for the second half of drilling—something that also did not happen, perhaps because of understaffing at Town.

The reviewers observed that they had found "no show stoppers" in the plan, suggested integrating the "keeper" option into the project's objectives, and asserted that "all major risk [*sic*] are addressed and mitigation developed."[32]

"Risk" clearly meant the same thing to the reviewers as it did to those who produced the register: risks to the timely and economical completion of the project, not to the *Horizon,* the lives of those aboard, or the ecology of the Gulf. Distance to The Edge was no longer a topic of interest.

CHAPTER FIVE

The Struggle with Macondo

THE STRUGGLE WITH MACONDO unfolded in two phases: during exploratory drilling, and during the production tail phase, when the *Horizon* crew would insert production casing into the well and attempt to use cement to isolate the bottom of the casing from hydrocarbons in the pay zone.

During drilling, the struggle centered on two aspects of a measurement called the *drilling margin*. The first was the density of mud or other drilling fluid required to push against the shove of pore pressure. The second was the limit on push set by the formation's fracture pressure. Too little push from the drilling mud and the well would kick: hydrocarbons from the formation would begin to enter the well, which could lead to a blowout. Too much push from the mud and the formation could break down and generate lost returns, which in turn may reduce the amount of push to a dangerous degree.

The range of "right amounts" of push is the drilling margin. It is a measure of the reserve push a driller has available to counter any unanticipated increases in shove, and as such is a basic measure of distance to The Edge.

Lost returns are classified as well-control incidents because they are dangerous. That danger is one reason regulations then and now—as well as accepted industry practices—mandate that drillers maintain a drilling margin of at least 0.5 pounds per gallon, or 0.3 ppg with prior approval from MMS.[1]

Drilling margin was a central issue during exploratory drilling of Macondo. An expert witness for the plaintiffs in MDL 2179 alleged that the well team mismanaged drilling margin, misread tests of fracture pressure, submitted optimistically incorrect reports to MMS, and at times drilled with no margin at all—in effect, tiptoeing along The Edge and trusting to luck.[2]

BP made a limited effort to refute these allegations during the trial by arguing that the Macondo team had followed the letter of federal regulations and thus was not technically in violation.[3] Given the refusal of key figures to testify under oath, we will probably never know whether this attitude reflected a conscious attempt at deception or go fever.

One clear outcome was that drillers sustained significant lost returns during the exploratory phase, pumping slightly over 16,000 barrels of mud irretrievably into the formations around Macondo.[4] As a result, concern with fracture pressure heavily influenced the planning and execution of the production tail phase. This concern, though, did not result in a decision to fill the bottom of the well with cement and abandon it, and drill a second production well using a design that took into account what the team had learned about the formation. Instead the team made a series of compromises that combined with go fever to cause the production tail effort to end with a tragic trip over The Edge.

THE MACONDO DESIGN

Before installing the production casing, the crew had to drill and build the external structure of the well. This process begins with a *wellhead*—a large fitting that extends above and below the *mud line,* or surface of the seabed. The fitting is cemented in place to form a firm foundation for the BOP and riser, and later the fittings that will connect the well to extraction facilities.

Progress after that occurs in stages, called *intervals*. At each interval, the crew drills deeper until it is close to running out of drilling margin and the formation is in danger of fracturing. At that point crew members run external casing—not to be confused with production casing—to reinforce the well. They attach this casing to the wellhead with a casing hanger at the top, and cement it in place at the bottom. Then, after testing the formation, they drill ahead into the next interval.[5]

Because each casing interval must slip inside the previous ones, the exterior of the well gets smaller and smaller in diameter as drilling proceeds, ending up looking like an inverted telescope, as shown in Figure 5.1. (This diagram, like all others in this book, radically compresses the vertical dimension. If a model of Macondo were made with the final diameter of the well the thickness of a human hair, the distance from the rig to the bottom of the well would be eleven feet.)

The design of a well is captured in a casing schematic, which shows the depths to which the crew will run each interval of external casing. The initial casing schematic for Macondo was the result of a cooperative effort between the senior drilling engineer in BP Town and a group of internal geologic consultants called the TIGER team. The geologists used data from the 1998 and 2003 surveys, along with records of nearby wells, to predict when drilling margin would decrease to the point where the crew would need to install casing to keep the well from fracturing.

Any such plan is subject to modification as measurements during drilling allow geologists to refine their forecasts of pore pressure. In an exploratory well, as Macondo was before its last few days, these modifications can be significant: the surest test of what is down in the formation is to drill into it.

The primary target of the well was a geologic zone known as M56— 18,400 feet below the mud line. The initial objective was to drill to 20,200 feet below the mud line, to test zones below M56, as noted in BP's May 2009 application to MMS for a permit to drill Macondo. By the time the *Marianas* was scheduled to start drilling the well, the design had evolved into the one in the schematic shown in Figure 5.2.

The design described a well that went to a revised depth of 19,650 feet below the mud line in seven intervals—the three topmost being part of the wellhead and installed as a unit. The major interval was that of the 16-inch-diameter casing, which ran from the wellhead down to a planned 12,500 feet. This casing was designed to deal with a difficulty forecast in the risk register and the peer review. The initial survey had discovered that these depths contained depleted sand: porous sand with low pore pressure, or shove, to work against the push of the mud, and therefore highly prone to lost returns.

The design incorporated three rupture discs in the 16-inch section of casing, to prevent the 7,500-foot-long piece from collapsing from internal or external pressures called annular pressure buildup.[6] This buildup occurs after the well is in production, and results from expansion of fluids when

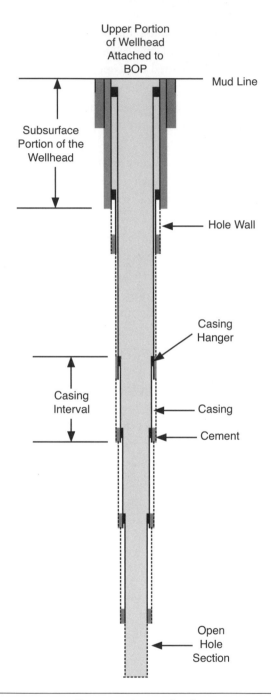

Figure 5.1. A typical well.

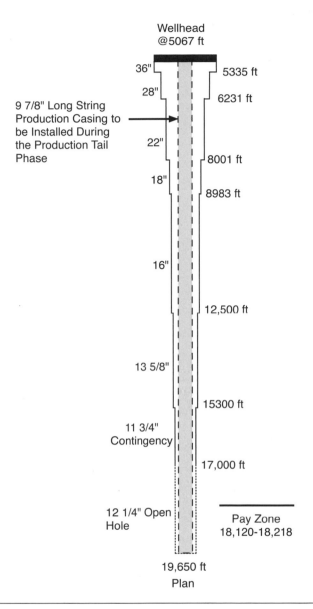

Figure 5.2. Macondo as planned.

they are heated by the hot oil flowing out of the pay zone and up the production casing.

The design also added an 11¾-inch contingency casing to the schematic originally submitted to MMS.[7] Evidence introduced during MDL 2179 included three casing schematics for the well.[8] Nothing in the evidence explains why BP Town made these changes.

An expert in well design who reviewed this design for the National Commission stated, "The initial design was adequate for an exploration well."

The expert was then asked about later designs for the production tail: "I felt that they [the designs] were deficient in detail, especially in light of the fact that by this point it had become fairly apparent that this well was going to be completed as a production well."[9]

CONSTRUCTING MACONDO

In October 2009, BP directed the *Marianas* to proceed to Macondo and initiate, or *spud,* the well. This was two months later than the plan called for—a delay caused by problems with the *Marianas*'s BOP and the need to remediate a bad cement job at the vessel's previous assignment, a well called Na Kika.[10] This was the first of multiple delays that would ripple through rig schedules and back the *Horizon*'s crew up against the Kaskida deadline.

At noon on October 6, the crew of the *Marianas* lowered a high-pressure jetting assembly to the mud line at the Macondo site and began hydraulic excavation of a 36-inch hole, using water-based mud to resist any unexpected shallow gas. By 10 p.m. crew members had the hole they wanted and had installed the beginning of the wellhead.

Two weeks later, after they had finished installing the wellhead, they ran into their first well-control incident, losing 431 barrels of mud into the formation over several days and taking a kick. This was the first of four instances regarding which the plaintiffs' expert witness accused BP of drilling ahead with less than the mandated drilling margin, and of being evasive in reporting to MMS the drilling margins they were using.[11]

This first well-control incident appeared to cause concern in BP Town—but not about safety. Rather, in an email exchange BP Town employees expressed concern about the "letter of the law" and discussed whether to seek clarification on how MMS wanted drilling margin measured. A drilling engineer on the *Marianas* team closed off the discussion by saying "maybe this is best left ambiguous with MMS."[12]

At some point an unidentified BP Town employee reviewed the resulting dilemma in terms of risk and reward. The crew could run external casing immediately, to strengthen the well, but in so doing would "potentially sacrifice hole diameter" in the pay zone. That is, the "inverted telescope" design of the casing could make the well too narrow to be useful, although it would prevent a "potentially uncontrollable" well-control event. If the crew drilled ahead without installing external casing at this point, on the other hand, it risked an uncontrollable event.[13] The crew drilled ahead without running external casing.

Later testimony by the pore pressure specialist on the rig suggested that the risk–reward analysis overestimated the degree of risk at this point in the project.[14] Whether that was or was not the case, the Macondo team thought the risk was real and continued to drill anyway. No crew member on the *Marianas* appears to have objected.

After drilling ahead, the crew cemented the bottom of the well hole to prepare for hurricane season. Hurricane Ida was active for seven days in November 2009. It spun up south of Nicaragua in the Caribbean Sea, clipped the coast of that country, and moved north past the Yucatan Peninsula, losing and regaining strength.

The tempest was at the transition between hurricane and tropical storm strength when it rolled over the *Marianas,* which was anchored over Macondo, her riser unlatched from the well and her crew flown safely ashore. The force of the storm dragged the *Marianas,* anchors and all, 250 feet off station. This was not surprising considering that forty-five miles to the southwest, over the Kodiak well, the *Deepwater Horizon* was reporting 71-knot winds.[15]

At Kodiak, the swells generated by Ida were forecast to be too great to allow proper functioning of a mechanism that automatically adjusts the length of the *Horizon*'s riser, and her crew had taken the precaution of activating the emergency disconnect system (EDS). The EDS automatically shut in the well and separated the two halves of the BOP, effectively unlatching the rig from the well.[16] Six months later and under far worse circumstances, that system would fail, with dire consequences for the rig and the Gulf.

THE *MARIANAS* LEAVES

When crew members evacuated the *Marianas,* they set her ballast tanks so she rode at her lower drilling depth, presumably to reduce the chances that the storm would cause her to drag her anchors—hurricanes have been known to move anchored rigs seventy miles. Possibly because of that, she sustained damage to the electrical conduits that affected her anchor winches and drilling equipment. After failed attempts at repair, on Thanksgiving Day the *Marianas* crew capped the well with a cement plug, pulled up her riser and BOP, and had the rig towed off station and into dry dock.[17] The ocean above Macondo was empty.

Meanwhile administrative employees in BP Town were occupied with processing an authorization to spend $27 million from the budget that had been approved for Macondo by BP headquarters. BP had negotiated with Anadarko Petroleum to enable that company to purchase 25 percent of the well, and another company, MOEX Offshore 2007, to buy 10 percent.

BP's share of the budgeted cost of the well was $20 million,[18] which covered the remainder of the lease on the *Marianas.* Because the lease was a contractual obligation, the three corporations signed off on the agreement in three days without objection or special review.[19] BP Town then submitted an application to MMS to replace the *Marianas* with the *Horizon* on Macondo, which the agency approved without difficulty.[20]

During this ninety-day lull, no one seems to have pushed ahead with engineering specifications for Macondo—not even the "optimized work instructions and logistics planning" that the review team had recommended almost six months earlier. In retrospect this was not surprising, for roughly coincident with the departure of the *Marianas,* Town was beginning to lose its ability to effectively control the project, never to regain it.

From roughly the time the *Marianas* left Macondo to the *Deepwater Horizon* blowout, Town operated almost completely isolated from the rest of BP. It could just as well have been a Silicon Valley startup, left to its own devices regarding how to expend limited resources supplied by a venture capital firm—in this case upper BP management.

This isolation was partly by design, as it corresponded to BP's decentralized model, and partly the result of personnel decisions made by the Gulf of Mexico SPU and BP corporate. Figure 5.3 shows BP Town's organization at this time.

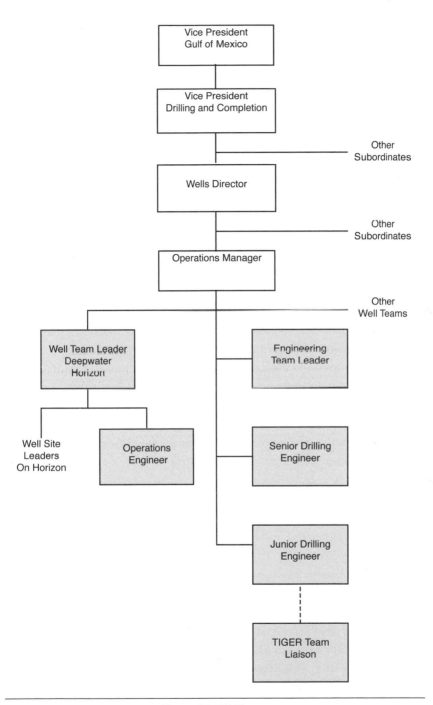

Figure 5.3. BP Town.

The title "vice president of Gulf of Mexico" was an informal one BP presented to the outside world. Inside the corporation this position was known as *SPU leader.*

Reporting to the SPU leader was the SPU's *vice president for drilling and completion,* who oversaw efforts to design and construct wells in the Gulf. Below this position was a wells manager who acted as chief operating officer, concerned with day-to-day activities. One more level down was the de facto head of Town, called an *operations manager:* "operations" was the term BP used for "well construction."

At the end of 2009 BP corporate replaced the three executives above the operations manager. The vice president for the Gulf of Mexico transferred out of that position in November, and the vice president for drilling and completions left BP a month later. BP corporate announced his departure as voluntary, but in fact he had been abruptly dismissed.[21] He was and remains well respected in the industry as a strong advocate for safety, but there is no clear evidence that such advocacy was a factor. He was apparently the victim of upper-level corporate politics.

His deputy, the wells director, voluntarily transferred at the same time, and the operations manager for BP Town began transitioning to a new job in London in February 2010, just as the *Horizon* arrived at Macondo. That transition left the position of operations manager effectively unfilled until shortly before the blowout. The transition also left Town effectively unsupervised by upper-level managers until the new vice president for drilling and completions decided to visit the *Horizon* the day Macondo blew out.

A TROUBLED TOWN

Town was composed of two teams, the engineering team and the well team. The engineering team included a team leader who supervised one senior and one junior drilling engineer. The senior drilling engineer had thirty years of experience and the junior five; Macondo was his third or fourth well.[22] The two of them were responsible for designing the well.

The two drilling engineers worked with a liaison geologist from the TIGER team, a distributed collection of geologists and petrophysicists who advised multiple activities in Town. The liaison geologist was also a junior professional, and was assigned a senior geologist to act as mentor and fill in when the liaison geologist was absent.[23] Email evidence indicates that the senior geologist devoted only a small percentage of his time to overseeing the junior geologist's work.

The well team was responsible for conveying to the drilling team at Macondo the procedures that would be used to construct the well, coordinating with Transocean, and ensuring that the proper materials and tools were available. The well team leader supervised four BP *well site leaders* (WSLs) assigned to the *Horizon*,[24] who worked a rotation that put two of them on the rig at any given time. He was supported by an operations engineer, who acted as liaison to the engineering team.

Tension between an engineering group and an on-site team is inherent in any form of construction: one group works in an idealized world, while the other deals with the messy reality of physical materials and the relentless march of deadlines. The leaders of the two groups must work well together to aggregate information each receives.

That was not the case at BP Town during the bulk of the Macondo project. The engineering team leader and the well team leader had a difficult and contentious personal relationship.[25] To make matters worse, when the operations manager was absent, the engineering and well team leaders alternated as acting operations manager. This arrangement meant that the two candidates for promotion to operations manager took turns supervising each other—a scheme that did little to reduce the apparent friction between them.

When asked if that situation could have affected safety, the fired vice president responded: "Any relaxation or distraction could potentially lead to a significant incident."[26]

The team in BP Town working full-time on Macondo included just five professionals, which left little or no resilience in the face of absences owing to other projects or personal concerns. The largest number of professionals in BP Town working on the project at any one time was twelve. As BP Town began turning its attention to Kaskida in late March 2010, the senior drilling engineer gradually disappeared from the Macondo project and the junior drilling engineer assumed more responsibility: "[The senior drilling engineer] acknowledged that [the junior drilling engineer] was not the lead drilling engineer 'on paper' for the Macondo well, but . . . in practice, he functioned that way. If he had issues or concerns he elevated them to [the senior drilling engineer] or others."[27]

During the last two days of the project, Town consisted essentially of the junior drilling engineer and the well team leader, who made decisions without consulting or in some cases even informing other members of the team. This contrasts with oil companies such as Shell, one of whose senior technologists told the National Commission that the company assigned fifteen to twenty professionals to each well.[28]

THE *HORIZON* ARRIVES

The *Horizon* was still on the Kodiak well when Ida hit. The crew had to finish up its work there and perform much-needed maintenance, so was unable to begin the eleven-hour transit along the forty nautical miles to Macondo until midnight on January 21, 2010.

Once at Macondo, the crew conducted various tests to ensure that its equipment was seaworthy, and then spent six days performing maintenance on the BOP, checking hoses and connections and replacing seals and gaskets. The one thing they did not do was replace the batteries essential to the BOP's emergency disconnect system, designed to automatically separate the *Horizon* from the well in an emergency.

Transocean called its approach to such work "condition-based maintenance": crews would not perform maintenance until the equipment required it. The chief mechanic on the *Horizon* called this policy "run it until it breaks."[29] A different philosophy—periodic maintenance—would have entailed replacing the batteries at fixed intervals even if they passed a test. The batteries would become a factor in the failure of the EDS to operate after the blowout.

The Transocean crew did swap out a component on one of the two BOP control pods—a component that was shown after the explosion to have been installed backwards. This was yet another example of maintenance shortcomings on the part of Transocean.[30]

The crew then spent five days installing the BOP and connecting the riser to it. In the middle of that process, the crew committed a series of errors while pressure-testing lines on the BOP. Those errors revealed a lack of crew coordination, and were the sort of thing that could damage a well. This prompted an investigation by the Transocean office in Houston, and a warning letter to a member of the drilling crew.[31] However, no one recorded this event on the consolidated log of rig activity, and there is no evidence that BP took any action as a result.[32] This incident suggests that the crew of the

Horizon had a tendency toward complacency when not actually "making hole"—complacency that would also appear when the crew performed dangerous activities during the rig's last hours.

Finally, around dinnertime on February 11, the crew made the first progress on the well in more than ninety days. At that point specialists in taking downhole measurements and assessing pore pressure joined the crew on board the *Horizon,* carrying equipment to directly measure what was going on down in the well.

THE LAST CONTINGENCY STRING

Despite a resumption of drilling, the *Horizon* was no luckier than the *Marianas* in coping with the Macondo formation. Six days and 3,000 feet in, the crew began battling lost returns in the open hole that was to take the planned 16-inch casing. The crew struggled with this for eighteen days, losing 4,700 barrels of expensive drilling mud while pumping in "pills" composed of viscous *loss control material* (LCM) to plug pores in the sand. The crew finally succeeded in installing the 16-inch casing.

The crew then drilled ahead for fewer than 1,000 feet and took a kick. When the crew tried to pull the drill bit and pipe out of the well, the pipe got stuck, and an expensive measuring tool was lost. The stuck pipe forced the crew to install a sidetrack: that is, to plug the hole with cement and use a directional drill to bore at an angle and then down, producing a new hole parallel to the original.

This exercise cost crew members another fourteen days before they again reached the depth at which they had taken the kick. More significantly, the sidetrack forced them to install a 13⅝-inch casing interval that was 2,000 feet short of the depth planned for that interval. That, in turn, meant that the crew had to start installing the 11⅞-inch contingency casing. At this point one of the measurement specialists on board emailed Town, "The mood is 'please god lets get this well behind us.' "[33]

The kick prompted a bout of soul-searching among members of the TIGER team, which recommended closer coordination between the drilling crew and specialists on the rig who were monitoring and forecasting pore pressure. The tone of the recommendations—like almost everything produced by Town—emphasized project risks and made no mention of safety risk.[34] When asked later, the senior geologist who initiated the recommendations stated that there was no need to mention safety explicitly because "by virtue of this e-mail being written under the BP Code of Conduct and the HSE [health, safety, and environment] principles, I've implied it."[35]

The mood at the working level was evidently more apprehensive. As the crew approached the 11⅞ external casing interval, the liaison geologist on the TIGER team emailed the TIGER team leader, "If they want to push this next hole-section to TD [total depth], it'll all be in God's hands."[36]

Macondo was not done with the team yet. On the last weekend in March, when employees in BP Town were attending a kickoff meeting on Kaskida, further losses into the formation forced the *Horizon* crew to install a new interval of external casing of even smaller diameter: 9⅞ inches. The implications of this move were profound. The original design for the production casing had envisioned a single 9⅞-inch-diameter *long string* running inside the well from the wellhead to the bottom. This approach was no longer possible, given that the external casing was now that same diameter.

The lower end of the production casing would now have to be 7 inches—nearly the minimum feasible for extracting oil. For some undocumented reason Town decided to use a *tapered long string*, one that was 9⅞ in its upper portion and 7 inches in its lower portion, with a fitting called a *crossover* joining the two portions, rather than the single 9⅞-inch-diameter long string they originally planned, or other options such as a *liner*, a short length of 7-inch-diameter casing attached near the bottom of the well. A tapered long string is a relatively rare configuration and raises questions about whether the junior drilling engineer who designed it had the experience to anticipate possible problems.

THE FINAL PUSH TO THE PAY ZONE

After the drill team lost 700 more barrels of mud, April brought another bout of lost returns, as the sands of Macondo absorbed another 3,000 barrels. The crew used more LCM pills to plug holes in the formation, expending five more days without making progress toward the planned drilling depth. The lost returns were so severe that crew members mixed a double quantity of LCM in anticipation of future needs.[37]

The crew then took steps to provide enough distance between the planned production casing and the formation to accommodate a thickness of cement that would resist pore pressure from the pay zone. In the section of the well hole below the 9⅞-inch casing, the crew used an "under reamer"—a tool behind the main drill bit with retractable and extendable cutters. The crew retracted the cutters to slide the tool through the casing, and then expanded the cutters to open the hole to 9⅞ inches.

On April 7 the senior drilling engineer submitted the one and only management-of-change document prepared during the exploratory phase—a document recording the end of that phase.[38] Two days later the crew drilled a final 100 feet and lost another eighty-one barrels of mud. The crew was now drilling in a formation where the mud weight required to push against the shove of pore pressure greatly exceeded the strength of the formation. Primary well control was no longer possible. In the words of one TIGER team member: "We had simply run out of drilling margin. At this point it became a well safety and integrity issue."[39]

The plaintiff's expert during the trial concluded that the team had encountered "well safety and integrity issues" several times previously, by using less than the mandated drilling margin.[40] But this time they had struck oil. Town employees passed word up BP's management chain that Macondo was officially a "discovery."[41]

THE SHAPE OF THE WELL

Although the crew had drilled 18,320 feet below the rig floor into a formation that was both delicate and dangerous, for the moment Macondo was stable.

Figure 5.4 shows the shape of the well as planned in August 2009 and as it was finally drilled on April 9, 2010.[42] The figure notes the depths at which well-control incidents occurred, and the sections of the well that the plaintiffs' expert witness asserted had been drilled in an unsafe manner. BP did not provide a significant rebuttal to that testimony during the trial.

At this point a BP petrophysicist joined the geologists on the rig and used a specialized "wireline" tool to measure the characteristics of the last thousand feet or so.[43] However, the shape of the well prevented the tool from reaching the bottom, leaving the last eighty feet unmeasured. This meant that BP Town did not know the exact depth of the well—a gap that would complicate installation of the production casing.

The measurement included a *caliper log,* during which feelers measure the diameter of the hole at specific intervals. This log showed that the bottom

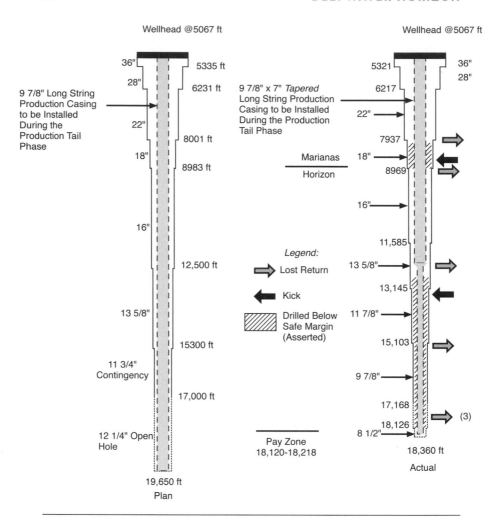

Figure 5.4. Evolution of the Macondo design.

of the hole, at least as far down as measured, was seriously washed out—eroded by the circulating drilling mud. The well diameter in some places was more than 3 inches larger than the intended 9⅞ inches, which would affect efforts to cement the bottom of the production casing.[44] No evidence shows who on the Macondo team may have received this information.

Using information from the same wireline run, a BP petrophysicist classified a thin stratum above Macondo's pay zone, called M57B, as sandstone containing pure brine. After the blowout, analysts found out that M57B actually contained a mixture of brine and hydrocarbon gas, with an estimated pore pressure about 700 psi higher than the pressure engineers were using to plan the installation of the production casing.[45]

If engineers use too low a pore pressure to create such plans, they may specify the use of drilling mud that is not heavy enough to withstand the pore pressure. The result can be a kick, and even a blowout. The plaintiffs' expert witness described the petrophysicist's error in a report whose main conclusions BP did not challenge in court.[46]

Email evidence supports the expert witness's contention that the petrophysicist spent just twenty-seven minutes reviewing a printout of the wireline measurements and did not enter them into software for analysis.[47] What's more, there is no evidence that the TIGER team or other employees at BP Town had any mechanism or desire to review her work. While it is unclear what effect the increased pore pressure may have had, the mistake shows every appearance of being another rushed activity by a bare-bones organization. That organization was now not only coping with difficulties at Macondo and transitioning to Kaskida but also completing its reorganization.

THE UNSEEN EDGE

By April 14 the wireline crew had departed the *Horizon,* taking their tools with them. Remaining crew members were testing equipment and performing maintenance and housekeeping tasks while waiting for the instructions and equipment needed to finish their work at the well.

The *Horizon* had, by all appearances, won the struggle with Macondo, and once again lived up to its reputation as one of the best rigs in Transocean and possibly the Gulf at drilling exploratory wells. In the glow of this success, neither rig nor Town paused to consider how close installation of the production casing would take them to The Edge.

The information they could have used to estimate their distance from The Edge was highly disaggregated. In Town, knowledge and appreciation of the fragility of the formation was concentrated in the TIGER team. Risks that may have arisen from the switch from a unitary to a tapered production string were known mainly to the engineering group.

What's more, two critical plans for completing work on the well were not yet finished. These plans would specify procedures for installing a cement barrier at the bottom of the well, capping the well until actual production began, and displacing heavy drilling mud in the riser with lighter seawater. The latter activity would directly affect well control during placement of the production casing.

The *Horizon* crew was in transition between drilling an exploratory and appraisal well—an activity with which they were familiar—and setting the production casing, which they had done only a handful of times. In fact, the crew had not even been told that it would be performing that task, as the supervisor of all drillers on the *Horizon* explained to a plaintiffs' lawyer:

> Q. Did the well plan include a plan to set production casing at Macondo?
> A. To the best of my knowledge, I don't remember that being a part of the plan.[48]

Rather than an integrated plan, BP Town sent the rig crew a disaggregated sequence of tasks. Town also sent several of those tasks at the last moment, and imposed some over the crew's objections. The Macondo team was about to lose control at every important level of the system: in Town, on the *Horizon,* and down in the depths of the well.

CHAPTER SIX

Planning the Production Tail Phase

THE ACTIVITY that BP called the "production tail" is more commonly known in the industry as *temporary abandonment*—a set of steps in which a crew puts a well in a stable state so that it may be left alone for a period of time, after which the same or a different rig and crew can resume work on it.

A crew that intends to abandon a well is somewhat in the position of an individual who has shaken an open bottle of soda while holding a thumb over the top, and who now faces the problem of replacing the thumb with a cap without allowing any soda to escape. In an abandonment, the thumb is the combined push of the mud in the riser and the well, which holds back the shove of the pore pressure, backed up by the barriers of the BOP. The cap is the reduced push of mud left in the well after the crew has detached the riser and BOP, augmented by cement barriers that the drill crew sets by pumping cement into the drill pipe.

Planning for the abandonment of Macondo was extremely complex. The fundamental source of that complexity was a phenomenon well known to systems engineers: the number of potential pairwise interactions among a set of N elements grows as N times $N-1$, divided by 2. That means that if there are two elements in the set, there is one potential interaction; if there are five elements, there are ten potential interactions; ten elements, and there are forty-five; and so forth. If the interactions are more complex, such as when more than two things combine, the number is larger. Every potential interaction does not usually become an actual one, but adding elements to a set means that complexity grows much more rapidly than ordinary intuition would expect.

In the case of Macondo, the elements of the set were the tasks that BP Town included in the abandonment activity, some of which were clearly required and others of which were more or less arbitrarily added. This inherent complexity—and attendant confusion—was worsened by BP Town's habit of not submitting to management decisions about change or any other process that reviewed each decision in the context of others. The complexity and confusion were also worsened by the manner in which BP passed responsibility for defining the tasks entailed in abandonment from individual to individual—sometimes crossing corporate boundaries—and captured them in multiple documents of different formats and levels of detail, resulting in a collection of material that was a plan in name only.

A witness testifying for BP admitted its lack of overall leadership and coordination of the process of defining this "plan:"[1]

> Q. And as you sit here today on behalf of BP, you have no knowledge, have seen no documents as to exactly how this Abandonment Plan was authored, who authored it, or who reviewed it to determine whether or not Risk Assessment or any changes needed to be made.
>
> [Response omitted from transcript.]
>
> Q. Correct?
>
> A. Correct.[2]

A compressed schedule that had the crew starting some tasks before related tasks were fully defined added yet more complexity to the abandonment process. Anyone who seeks to understand the *Horizon*'s path to The Edge must understand how the approach of making one disjointed decision at a time created that complexity.

Our approach to gaining that understanding proceeds in stages. In this chapter, we first describe the way a well looks after a "simple" abandonment—the kind the *Horizon* crew was accustomed to performing. We then present the way BP intended Macondo to look after the crew had installed the production casing, which we have reconstructed using evidence from the trial. That configuration represents, in effect, the specification BP Town never produced in consolidated form. From this we enumerate the individual tasks required to produce that configuration.

Later chapters will describe the evolution of the plans for three significant tasks: designing and fabricating the production casing, cementing it in place, and testing the well for integrity while displacing the heavy mud pushing against the shove of pore pressure with lighter seawater. We then will show how those tasks influenced each other, and why they required procedures whose risks were not visible either to the individuals who defined the tasks or those charged with accomplishing them. Those risks were the direct result of BP's disaggregated decision making and lack of contextual review.

SIMPLE ABANDONMENT

Abandonments are routine in the Gulf. In 2010 the Associated Press estimated that the Gulf of Mexico had 27,000 abandoned wells—nearly every one of which had undergone a process similar to the one we describe.[3] However, although the process of abandoning a well is common, it is also dangerous and delicate, because it involves deliberately reducing push.

The diagram in Figure 6.1 shows how a well that is "latched up" and monitored is transformed into one that is abandoned. The *Horizon* crew had performed this "simple" or "pure" abandonment dozens of times, and the crew members could perform their duties largely from memory—a familiarity that permitted them to operate without written procedures or formal checklists.

The first step in a simple abandonment is to seal off the hydrocarbon-bearing zones with bottom plugs composed of cement. There are usually at least two, one set in the formation above the zones, and one set at the bottom of the last casing interval. These are called balanced plugs: the crew sets them using hydrostatic forces, or a metal barrier underneath them, or both. The plugs impose no significant stress on the formation (and none at all in the case of the plug inside the last interval of casing string) and therefore can be set

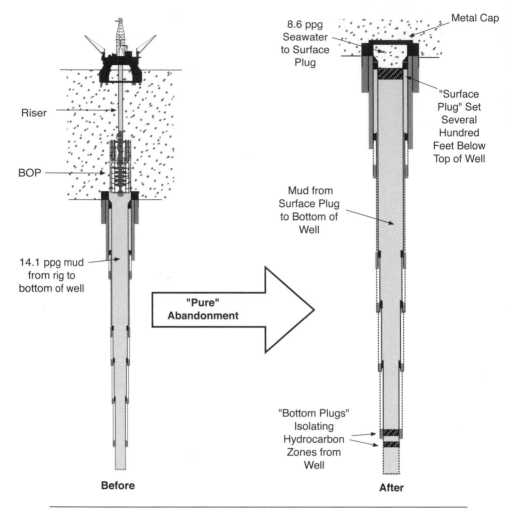

Figure 6.1. "Pure" abandonment.

in highly fragile formations. The crew then sets a top, or surface, plug some short distance below the wellhead, also as a balanced plug.

After the cement has set, the crew checks the integrity of the well using a negative test,[4] which determines whether the residual push from the trapped mud combined with the mechanical strength of the cement plugs can resist the shove of the highest pore pressure. If the well passes the test, the crew displaces mud in the riser by pumping seawater in from the bottom. That, in turn, enables the crew to pull the riser up onto the rig without losing mud into the Gulf. The crew then pulls up and stores the BOP, and the *Horizon* is ready to move to her next assignment.

The *Horizon*'s Negative Test

When a crew begins the process of abandoning a well, the well is a pressure vessel of unknown strength whose integrity the crew must assess before removing the riser and moving the rig away. As we noted in Chapter 1, in other process safety environments involving pressure vessels, such as oil refineries and chemical plants, crews test those vessels before loading them with hazardous material. The consequences of a test failure in those environments are therefore limited.

In contrast, the *Horizon* crew—like any crew involved in well construction—had to test the integrity of the well while exposed to pressurized and explosive hydrocarbon gas, and do so in a way that prevented a failed test from turning into a blowout, explosion, loss of life and equipment, and damage to the environment. Testing the integrity of the well entailed measuring the *seawater gradient:* the difference between the push produced by the heavy drilling mud and the lesser push from the lighter seawater that would replace it in the riser after the well passed the test.

At the time of Macondo, neither BP nor Transocean nor MMS had a specified procedure for doing a negative test; every crew devised its own. Nor did MMS require such a test, although that changed after Macondo. MMS mandated only a positive pressure test, wherein the crew would use the rig's pumps to increase pressure inside the well to 2,500 psi, as a crude check on the integrity of all the cement work in the well.[5]

Because MMS and BP left the details of the negative test to the crew, the various directives BP Town sent to the rig noted only where in a sequence of actions the crew was to perform the test, not how to perform it. A senior WSL assigned to the *Horizon,* who had conducted many such tests but was off duty at the time of the blowout, noted in his testimony:

> A. But many cases, we get our procedures to the rig, there's no—it's just a one-line sentence that says, perform negative test to seawater gradient. It's not unusual to see just, negative test—perform negative test.
>
> Q. But you told us yesterday that it's best to have the procedures in writing, correct?
>
> A. Yes. And normally, most cases, we take care of that at the rig.[6]

Assigning responsibility for planning this test to the crew could work if the crew is performing the temporary abandonment procedure noted above—the one with which the *Horizon* crew was familiar. However, when BP Town sent similar one-line directives to the *Horizon* crew in the context of the unfamiliar tasks entailed in completing the production tail, they led to fatal confusion.

No written BP procedures for performing negative tests on previous wells have survived. The only documentation that could shed light on the procedures the *Horizon*'s crew might have planned to use is an email sent from one crew member to another before such a test on Macondo on January 28.[7] That test was part of preparations to open up the well after the *Marianas* had abandoned it.

What we do have is a detailed procedure outlined by one of the *Horizon*'s senior WSLs in his deposition.[8] However, interviews and other testimony show that the crew did not follow that procedure on the evening of the blowout. In fact, survivors and outside experts have disagreed on how many negative tests the *Horizon* crew actually performed, whether one or another procedure actually constituted a negative test, and what constituted success or failure for some of them.[9] Given that, the term "negative test" is meaningless in terms of denoting a specific set of actions.

Figure 6.2 shows the *Horizon*'s negative test as described by the senior WSL—the test we must assume the crew had performed on most, if not all, of its previous abandonments. We will use that description in later chapters to note shortcomings in what the crew actually did.

As noted, the purpose of the test was to subject the cement plugs in the well to the amount of stress they would encounter when the riser and its push against pore pressure were no longer there—and to do so in a way that allowed the crew to shut in the well if the cement failed during the test.

The crew would conduct the test with the drill pipe lined up to the *cement unit*. That unit—separate from the drill floor—contained its own variable-speed, high-pressure pumps, a calibrated tank to measure volumes of fluid from the well, and a pressure gauge that produced a paper record during the test. The crew inserted a high-pressure safety valve in the line between the drill pipe and the cement unit. The crew could close this valve manually if the cement failed and the well took a kick.

The crew would begin the test by filling the drill pipe and a short depth of the well with seawater, and then closing the BOP's annular preventer to

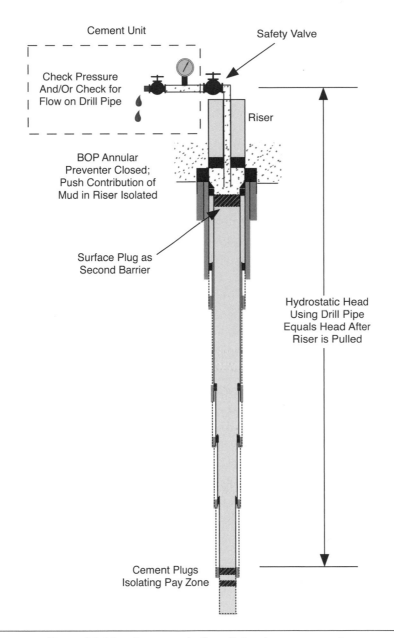

Figure 6.2. Negative test under "pure" abandonment scenario.

isolate the push of the mud in the riser from the well. That left the seawater gradient in the drill pipe to contribute to the total push. The WSL would then go from the drill floor to the cement unit to witness a check for a pre-calculated initial flow of seawater from the well to the tank.[10] He would then observe the drill pipe for thirty minutes to verify that the well was not continuing to flow—that is, that the cement barriers were preventing hydrocarbons from entering the bottom of the well and pushing seawater out the drill pipe.

The crew also reduced the risk entailed in a negative test by closing the annular preventer of the BOP, which isolates the riser from the well. If the well were to kick because of reduced push, the crew would detect the kick through the drill pipe. And because the well would, in effect, already be shut in, the crew could easily control the kick by closing the safety valve, and then take remedial action.

However, a negative test is tricky because it relies on a reading of zero flow from the well and no movement of the pressure gauge. That means the crew must set up the test carefully to ensure that nothing could produce a reading that suggests that the well is not flowing when in fact it is.

The senior WSL who outlined this procedure in his deposition also stated that in more than a hundred negative tests he had performed, only one had indicated that the well under test was faulty.[11] Such a high pass-fail ratio can—as it did on the *Horizon*—lead crew members to search for an explanation other than that they are dealing with an unsound well, if measures suggest that it is still flowing. This is a textbook definition of go fever.

The most serious shortcoming of the *Horizon*'s negative test was the absence of documented procedures for performing it. That was especially problematic because only two of the four WSLs assigned to the rig appear to have understood the test well enough to conduct it—and neither was on the rig when the crew performed the test on Macondo.

An Alternative Test

Another configuration for a negative test uses the *kill line,* the auxiliary line that runs down the riser and taps into the BOP below the annular preventers—the one that, as we described in Chapter 3, is normally used to pump heavier mud into the well after it has been shut in after a kick, in order to increase the amount of push.

The procedure used for this test is similar to that described above: displace mud in the kill line with seawater, close the annular to isolate the push from the mud in the riser, and observe the pressure and flow at the end of the line at the drill floor. The hydrostatics—and therefore the effectiveness—of both tests are identical.

The *Horizon* crew likely used this configuration so that they would not have had to reattach the cement unit to perform remedial cementing if the negative test indicated such was required.

ABANDONMENT WITH A PRODUCTION TAIL

Figure 6.3 shows the intended arrangement of the Macondo well after the crew had completed the production tail. The primary difference between this configuration and the much simpler one noted above is the production casing running from the bottom to the top of the well. (Chapter 9 will describe the convoluted process whereby BP Town arrived at this configuration.)

BP Town decided to set the upper concrete plug 8,367 feet below sea level, and about 3,300 feet into the well. This was an unusual depth for such a plug—one that no one associated with Macondo before or after the blowout had seen before. As noted, it was to be a balanced plug like those the crew had used in previous wells—except that the crew would plug the production casing as opposed to the whole well.[12]

The tasks implied by this configuration included running the production casing into the well, cementing it at the bottom, setting the lockdown sleeve and "surface plug," testing the integrity of the well, and displacing the mud out of the riser before pulling it and the BOP off the well. The nature of these tasks only partially constrained the order in which the crew could perform them. The loose constraints permitted BP Town to devise a physically possible but unsafe sequence—which is precisely what it ended up doing.

The constraints were that the crew had to run casing first and then cement it at the bottom. They had to displace mud with seawater last, and perform the associated negative test sometime after installing the casing and before displacement. They could install the lockdown sleeve and set the "surface plug" at almost any point in the sequence.

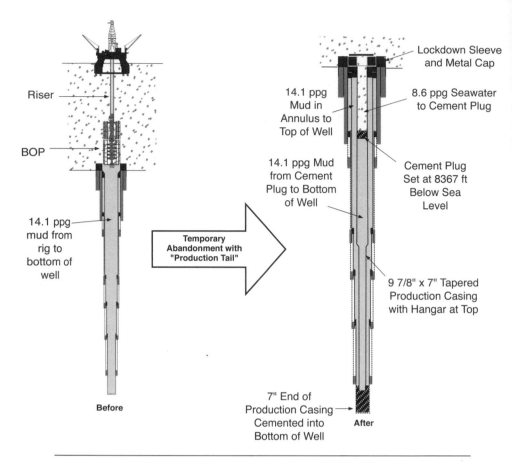

Figure 6.3. Abandonment with production tail.

Chapters 7, 8, and 9 explore the chaotic process BP Town used to plan for setting the production casing, cementing it to the bottom of the formation, and completing the combined activity of displacing drilling mud out of the riser and testing the integrity of the well. Although we explain these as separate projects, BP Town actually developed the plans for completing them concurrently—and even worked on plans for some while the crew was performing others. Figure 6.4 shows their schedules and the links between them.

The crew never did set the upper cement plug and the lockdown sleeve. However, the way BP Town planned those tasks influenced the displacement and testing plan in a way that caused it to violate one of the most basic precepts of well construction and well control.

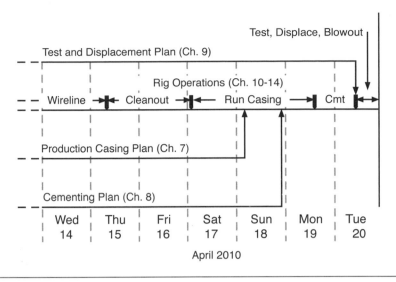

Figure 6.4. Relationship between planning and operations during the final week.

The addition of the production tail activity not only imposed a set of tasks on the crew of the *Horizon* that they had done infrequently, it changed two aspects of the negative test in equally unfamiliar ways. In the simple abandonment they had performed so many times before, the casing being tested was the external shell of the well, and it was tested to a seawater gradient running down to the top of the well. For Macondo they were required to test an interior production casing, and because the upper cement plug was set so deep, the seawater gradient being tested to ran down to 8,367 feet. This situation is shown in Figure 6.5.

The production casing test required that the well be displaced from below the annular preventer to 8,367 feet, so that there would be a continuous seawater gradient down the drill pipe to the depth at which the upper cement plug was to be set, thereby duplicating the stress the lower cement work would be subject to when the riser was pulled. BP Town vacillated as to whether the test was to be conducted using the drill pipe or the kill line, an uncertainty that was not resolved in favor of the kill line until the test was actually conducted. The kill line had a lower valve that was the possible source of a false reading that the test was a success—if closed it would block flow and pressure, and there is the possibility that it could have been closed without the crew being aware of it.

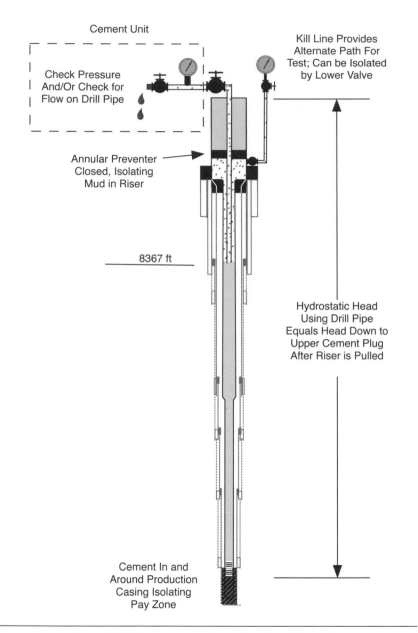

Cement Unit

Check Pressure And/Or Check for Flow on Drill Pipe

Kill Line Provides Alternate Path For Test; Can be Isolated by Lower Valve

Annular Preventer Closed, Isolating Mud in Riser

8367 ft

Hydrostatic Head Using Drill Pipe Equals Head Down to Upper Cement Plug After Riser is Pulled

Cement In and Around Production Casing Isolating Pay Zone

Figure 6.5. Negative test under production tail scenario.

All in all, the new configuration to be tested was sufficiently different from the one the crew was accustomed to that it called for a review and reformulation of the steps they had performed so often in the past. This was not done, and the result was confusion in the minds of those responsible for conducting it, confusion that combined with go fever to produce a fatal outcome.

CHAPTER SEVEN

The Production Casing Plan

PRODUCTION CASING, cement at the bottom of a well hole, and the well formation form a closely coupled subsystem whose elements strongly influence each other and the success of the subsystem—defined as the ability to isolate hydrocarbons from the well. The steps entailed in creating a closely coupled subsystem—especially when they involve disparate technologies—require contextual reviews to ensure that a decision that seems reasonable for one step does not adversely affect the others or cause an undesired event to emerge.

A contextual review of the plan for the production casing, conducted in this spirit, would have involved drilling engineers, cementing specialists, and geologists. However, as this chapter and others show, contextual reviews—including risk registers and management of change—were alien to BP Town's approach to well construction. That means no one individual, in Town or on the rig, likely had the view of the plan for the production casing that we present here.

The fact that BP reorganized just when key members of the Macondo team were facing unforeseen problems with components critical to ensuring the effective setting and cementing of the production casing worsened these shortcomings.

FROM LONG STRING TO TAPERED LONG STRING

From the very beginning, the plan for Macondo included a "long string" production casing running from the top to the bottom of the well. The original design included a casing with a single diameter of 9⅞ inches from top to bottom. This design did not survive the evolution of the external casing into an "inverted telescope."

As noted, during the March 8 kick, when a tool became stuck in the well, the crew had to perform a "sidetrack," use up the 11⅞-inch contingency string of external casing, and almost immediately run the 9⅞-inch external casing interval that BP Town had not planned for—an interval whose diameter led them to the tapered long-string design for the production casing.

Figure 7.1 shows the arrangement of the production casing, and how it was set in place and made ready for cementing.

The leftmost diagram shows the basic parts of the production casing: the *casing hanger,* which rests on the wellhead when the production casing is installed; the *crossover* between the 9⅞-inch and 7-inch diameters of the tapered long string; and the *shoe track,* the lower 189 feet of the production casing, which has a fitting called a *float collar* at its top and a perforated *shoe* at the bottom. At this point the float collar can be considered a perforated crosspiece in the casing; we will explain its full function later.

The middle diagram shows how the 600,000-pound, 2½-mile-long casing is attached to a *running tool* during final stages of installation. That tool, in turn, is connected to a string of heavyweight drill pipe called a *landing string* and lowered in place. In the rightmost diagram the production casing is landed, but a mechanical seal is left open to allow circulation down the production casing, through the shoe track, up the annulus between the production casing and the outer casings, and finally through the open seal and into the riser. This allows mud to flow into the riser as cement is pumped into the bottom of the well.

The tapered long string introduced a potential weakness at the crossover, and also complicated the upcoming job of cementing the casings to the well.

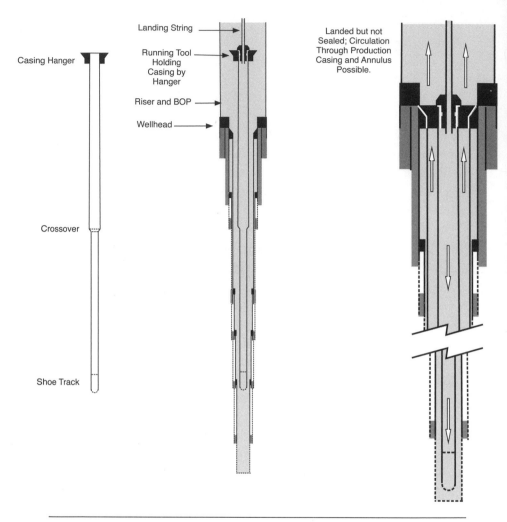

Figure 7.1. Configuration and installation of the production casing.

More significantly, BP did not have an inventory of 7-inch casing and associated fittings for the shoe track, and the lead time for manufacturing them ran into weeks. The BP procurement team was accordingly tasked with locating such material somewhere on the Gulf Coast. In another indication of the unusual nature of the design, the procurement team notified Macondo's junior drilling engineer that BP had no other need for 7-inch casing, and that Town would charge his project for it whether the crew used it or not.[1]

BP procurement finally located surplus unused casing and fittings in a storage facility belonging to Weatherford, which was under contract to BP to provide casing and fittings, and to assemble casing on shore and on the rig.[2] Although stored by Weatherford, the material belonged to an oil company called Nexen, and had evidently been purchased but not used for an appraisal well that finished up in early 2010. The material in inventory consisted of 6,000 feet of 7-inch casing and a set of fittings that could be used to make a production casing, as shown in Figure 7.1.

The casing material was sitting in two outdoor storage facilities and came with a strong disclaimer: "WARRANTY LIMITATION: GOODS ARE SOLD 'AS-IS', WHERE-IS AND WITH ALL FAULTS AND DEFECTS, IF ANY, KNOWN AND UNKNOWN, VISIBLE OR OUT OF SIGHT."[3]

BP had its experts check the documentation associated with the casing—a form called an MTR, which variously stands for "mill test report," "manufacturers test report," and "material test report." Such documents describe tests performed at some earlier time on the physical characteristics of the material. MTRs are notoriously unreliable, especially for stored casing: paperwork gets lost or mixed up, and inventories can get mingled as workers move casing around in the storage facility.

Stored casing may also be subject to "rack rot": internal corrosion of damp particulate matter, such as soil or fragments of the wooden racks on which the casing is stored, or external corrosion at the point of contact with the rack. This is a particular problem in Gulf coastal locations where the Macondo 7-inch casing was stored, and difficult to detect even with advanced inspection techniques.[4]

No one submitted evidence during the trial that explained how long the casing had been exposed to the elements, or the MTRs for this casing. There is also no evidence that anyone inspected the casing—given the haste evident in the procurement, it is unlikely that anyone did so. Thus the last 5,800 feet of the production casing, while unused, had been exposed to the elements for an unspecified period of time, and its condition was unknown at best and suspect at worst. There is also no evidence that anyone involved in the production of prior reports on the blowout was aware that the 7-inch casing and associated fittings used in Macondo were surplus.

BP did have an independent contractor inspect the fittings included in the purchase, as well as the threads on the sections of casing. The crossover had to be custom-made, and there is no evidence depicting how or even whether it was tested.

CEMENTING BASICS

A crew relies on the rig's circulating system to both control the cementing process and monitor changes in pressure and flow rates in the well. Those reveal whether the crew has performed the steps correctly.

The process begins with the well configured as shown in Figure 7.2. The BOP and the casing hanger seal are open to permit the mud displaced by the cement to flow back up through the riser and into the pits on the rig. A specialized running tool is attached to drill pipe and inserted into the casing hanger. This running tool holds two *wiper plugs* that will separate the top and bottom of the cement "package" from the drilling mud. The running tool and the wiper plugs have mechanisms in them that permit them to be released at the proper time.[5]

When cement is pumped down the drill pipe, it activates a mechanism that releases the bottom wiper plug, which then precedes the cement down the inside of the production casing, as shown in Figure 7.3.

The top wiper plug stays attached to the running tool while this is happening. From this point to the end of the cementing process, the returning mud provides pressure and flow signals to the circulating system to inform the crew how the process is proceeding. When the last of the cement has been pumped down the pipe, the crew causes the top wiper plug to be released and shifts to pumping mud again, which pushes the "package" of plugs and cement down the production casing, as shown in Figure 7.4.

When the "package" encounters the obstacle formed by the float collar, its movement stops and the pressure ruptures a diaphragm in the bottom wiper, permitting cement to flow into the portion of the shoe track below the float collar (Figure 7.5).

Pumping then continues, forcing the cement through the shoe and up into the annulus between the production casing and the formation until the top plug strikes the bottom one, sealing the hole in it. This step, called "bumping the plug," sends pressure and flow signals that tell the crew that the cementing is complete (Figure 7.6).

This sequence suggests the basic requirements for a "cementable" production casing. First and most obviously, the casing must be whole, without

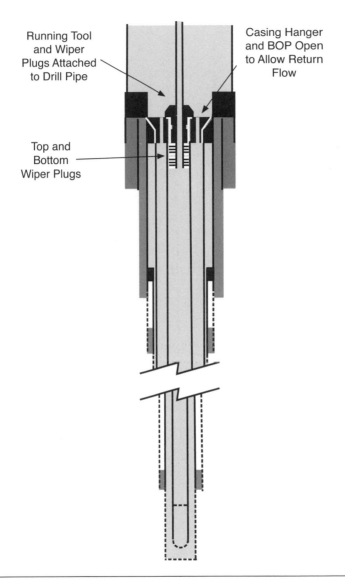

Running Tool
and Wiper
Plugs Attached
to Drill Pipe

Casing Hanger
and BOP Open
to Allow Return
Flow

Top and
Bottom
Wiper Plugs

Figure 7.2. Well with wiper plugs inserted.

cracks or breaches. The shoe track must be open so the cement can flow down into it and back up into the annulus between the production casing and the formation. Finally, the cement in the annulus must be of constant thickness all around the casing, or "channeling" can occur. Channeling allows hydro-carbon to flow along the thin spots, as shown in Figure 7.7.

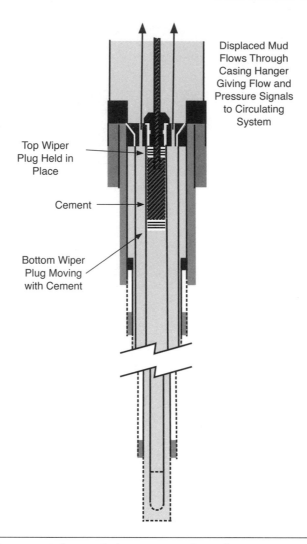

Displaced Mud
Flows Through
Casing Hanger
Giving Flow and
Pressure Signals
to Circulating
System

Top Wiper
Plug Held in
Place

Cement

Bottom Wiper
Plug Moving
with Cement

Figure 7.3. Releasing the bottom wiper plug and pumping cement.

The crew prevents channeling by placing centralizers along the casing, which have longitudinal springs to engage the sides of the formation and hold the production casing away from it. The lot of fittings that BP purchased from Nexen included six centralizers, as well as a float collar and two shoes, only one of which was needed for the production casing. The centralizers were in the form of centralizer subs: short lengths of 7-inch casing with centralizer springs attached, as shown in Figure 7.8.[6]

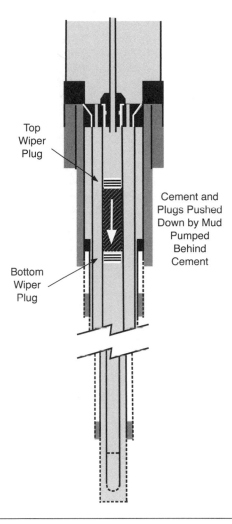

Top Wiper Plug

Cement and Plugs Pushed Down by Mud Pumped Behind Cement

Bottom Wiper Plug

Figure 7.4. Releasing the top wiper plug and pushing the "cement package" down the casing.

For Macondo, the use of subs restricted the engineers' ability to select the number and placement of the centralizers. Because they were subs, they could be located only between lengths of casing. What's more, the subs in the Nexen stock could accommodate only a 10¾-inch hole. Later wireline measurements showed that the zone to be cemented in Macondo varied erratically from 8½ to 12 inches in diameter, and because they only had six centralizers to work with, the engineers would have to place them with care.

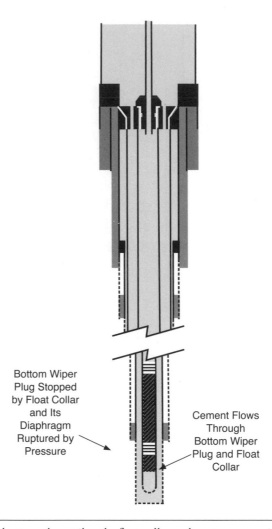

Bottom Wiper
Plug Stopped
by Float Collar
and Its
Diaphragm
Ruptured by
Pressure

Cement Flows
Through
Bottom Wiper
Plug and Float
Collar

Figure 7.5. The bottom plug strikes the float collar and ruptures, permitting cement to flow below it.

The engineers also had to use the type of float collar and shoe in that stock. Despite these challenges and others noted above—including the use of surplus material—the well team leader exercised his discretion not to subject the configuration of the production casing to any management-of-change process. He justified that decision by contending that the term "long string" covered both designs.[7]

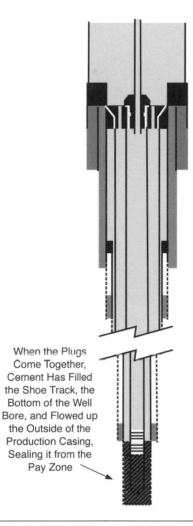

When the Plugs
Come Together,
Cement Has Filled
the Shoe Track, the
Bottom of the Well
Bore, and Flowed up
the Outside of the
Production Casing,
Sealing it from the
Pay Zone

Figure 7.6. "Bumping the plug": The top plug strikes the bottom one, signaling that the cement job is complete.

At this point, around April 12 and 13, the configuration of the production casing remained unsettled, the crew was still making wireline measurements of the formation, and the details of the cementing job were yet to be worked out. The job had no risk register, and no contextual reviews of significant documents had occurred. Instead, there was the major distraction of BP's reorganization.

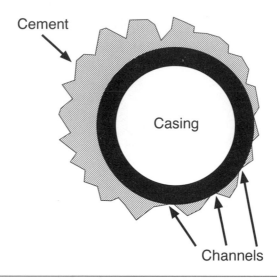

Figure 7.7. Channeling.

TOWN REORGANIZES

BP completed the two-week transition to the company's new organization during this period. The stated intent was to move BP closer to the functional model of other oil companies, which concentrated specialties in independent organizations. However, BP chose to phase in this transition—with the changes to Town being just the first step. The company would not reap the advantages of the new model until people holding positions such as chief driller and chief engineer had real power and appropriate reporting chains.

Like many half-measures, the reorganization did not succeed in increasing the influence of the specialties. Instead, it entailed a mass replacement of managers above Town with individuals unfamiliar with the projects and personnel under them. Within Town, the systemic effect was to further inhibit the aggregation of information by widening instead of closing the organizational gap between the two groups—engineering and operations—working on the well.

The original organization is shown in Figure 7.9, and the results of the transformation are depicted in Figure 7.10.

The junior drilling engineer was now in a new group with two levels of brand-new management, his old supervisor having been promoted and

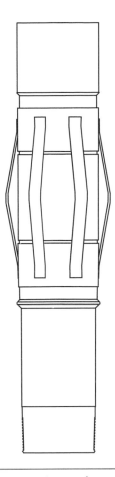

Figure 7.8. A centralizer sub.

moved over to the operations side of a more firmly split organization. The two sides of the new organization came together at a vice president whose background was in completions—the finishing off of already drilled wells. This individual was one of the visitors to the *Horizon* when Macondo blew out. His stated motive for the trip was that he wanted to see what a highly effective exploratory rig looked like.[8]

The vice president above him admitted that he did not know any of the members of the Macondo team, which individual was responsible for safety on the *Horizon*, or how difficult the geology of Macondo was.[9] The new wells manager was at a process safety school at MIT from early April until he was called back because of the explosion.[10]

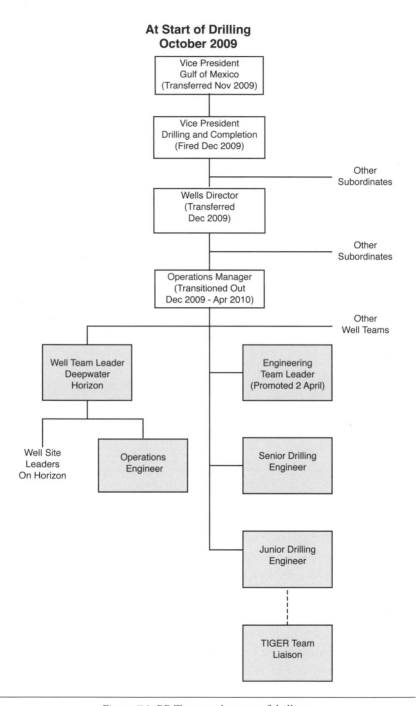

Figure 7.9. BP Town at the start of drilling.

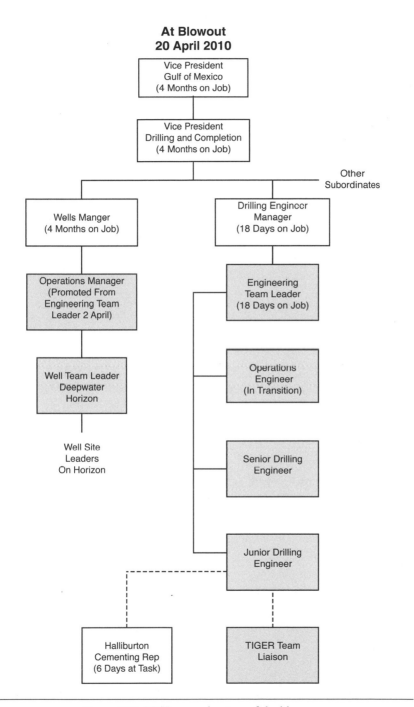

**At Blowout
20 April 2010**

Vice President
Gulf of Mexico
(4 Months on Job)

Vice President
Drilling and Completion
(4 Months on Job)

Other
Subordinates

Wells Manger
(4 Months on Job)

Drilling Engineer
Manager
(18 Days on Job)

Operations Manager
(Promoted From
Engineering Team
Leader 2 April)

Engineering
Team Leader
(18 Days on Job)

Well Team Leader
Deepwater
Horizon

Operations
Engineer
(In Transition)

Well Site
Leaders
On Horizon

Senior Drilling
Engineer

Junior Drilling
Engineer

Halliburton
Cementing Rep
(6 Days at Task)

TIGER Team
Liaison

Figure 7.10. BP Town at the time of the blowout.

The newly promoted operations manager had been occupied since April 9 with juggling rig schedules around Nile and Kaskida, and preparing a request to MMS to allow BP to suspend operations at Kaskida. He sent this request the day before the blowout, and it was still pending when the explosion occurred. Preparation of the request also distracted the junior drilling engineer away from Macondo.[11]

Adding to this turmoil was the temporary addition to the team in Town of a Halliburton representative who became the de facto leader of the cementing effort. He used a proprietary computer program called OptiCem, which accepted basic well parameters and cement characteristics as inputs. The program then calculated what the resulting cement job would look like, and how effective a barrier it would provide. The software also produced the procedures that would instruct the cementers on the rig how to mix the cement slurry and pump it into the well, when the final design was complete.

The Halliburton representative coordinated his work primarily with the junior drilling engineer, which added an extra drain on the time of an individual whom the reorganization had just deprived of any significant oversight, mentoring, or significant review of his decisions. His isolation increased a few days later when he—and the procedures for the production tail activity, which were his remaining responsibility—moved from Town to the rig.

THE CENTRALIZER DISTRACTION

On the afternoon of April 15, casing design, cementing, the well formation, and BP Town's brittle and newly shaken up organization came together in a disorganized exercise, one that probably contributed to BP Town's lack of consideration of how close the project was to The Edge. The junior drilling engineer was on his way to the rig, the senior drilling engineer was occupied with the results of a meeting on cementing whose import we will describe in Chapter 8, and the well team leader was home sick. There was no one in Town to handle any extra issue save the two engineering managers, who had been in their positions for just eighteen days.

The exercise began when the Halliburton representative sent out an email stating that his latest OptiCem software run—the first that used actual wireline measurements of the shape of the well hole—had shown that significant channeling would occur if the crew used only six centralizers on the production casing. He was evidently concerned enough about this that he repeated the message verbally to the engineering team leader, along with the software's recommendation that the crew install at least twenty-one centralizers on the production casing. Later investigations revealed that the representative had entered a pore pressure into his OptiCem software that was about 1,000 psi too high—the effect of this on the validity of the OptiCem output is unknown.

Unable to reach the well team leader, the engineering team leader conferred with the senior drilling engineer, and they jointly decided to attempt to get more centralizers to the rig. Weatherford had fifteen centralizers whose design was different from those on the rig: these slipped on, were held in place by rings called *stop collars,* and were affixed with epoxy. Weatherford promised to have them on a helicopter with a technician to install them by 6:30 the next morning. The engineering team leader then sent an email to the absent well team leader documenting what he had done.

Slightly earlier than that, the junior drilling engineer had requested a geodetic survey of the open-hole region. When emailed to him a short time later, this survey showed a remarkably vertical last section—just six-tenths of a degree out of plumb. He referred to this when replying to the Halliburton representative's email of concern, probably unaware that back in Town efforts were under way to obtain more centralizers: "We have 6 centralizers. We can run them in a row, spread out. or any combinations of the two. It's a vertical hole so hopefully the pipe stays centralized due to gravity. As far as changes, it's too late to get any more product to the rig, our only option is to rearrange placement of these centralizers."[12]

Attached to this email was a diagram showing where the crew was to place the six centralizers—the first such diagram in the trial evidence.

The next morning the Weatherford technician arrived at the airport to find only one of the two boxes containing his supplies. He got on the helicopter anyway and flew to the rig, where he attempted to locate his material, and received a promise that it was on its way by boat.

In Town, the well team leader returned to work to find the email about the additional centralizers. Around midday he canceled their installation on the casing, on the grounds that they were of a design that was prone to jamming when run into a well, and that the installation would take an extra ten hours.[13] There is no evidence that he was aware that the installation could not occur because essential parts were missing.

Sometime later that day the junior drilling engineer exchanged emails with a colleague on whether the vertical nature of the hole compensated for the small number of centralizers. The colleague noted:

> Even if the hole is perfectly straight, a straight piece of pipe even in tension will not seek the perfect center of the hole unless it has something to centralize it.
>
> But, who cares, it's done, end of story, will probably be fine and we'll get a good cement job. I would rather have to squeeze than get stuck above the WH. So [the well team leader] is right on the risk/reward equation.[14]

"Squeeze" is a reference to a remedial cement job, "WH" is the wellhead, and "risk/reward" refers to the concern that the centralizer springs could get caught as the crew ran production casing into the well—a concern that evidently overrode any worries about a bad cement job. The junior drilling engineer responded with a revised placement based on the wireline measurements of the hole, and there the matter stood until the production casing was run.[15]

This episode served only to divert the attention of key personnel, waste a day of an already compressed schedule, and delay any melding of BP Town into a functioning team. The newly appointed engineering manager who initiated the episode felt moved to apologize to his subordinates for his decisions.[16]

As Chapter 8 will show, the topic of centralization occupied analysis and testimony after the explosion that could have been better spent elsewhere. The real question was not how many centralizers it would take to successfully cement the production casing to the bottom of Macondo with foamed cement. The real question—raised only by the senior drilling engineer before the fact, and dismissed by his superior and the well team leader—was whether the bottom of Macondo could be successfully cemented with foamed cement at all.

CHAPTER EIGHT

The Cementing Plan

OIL COMPANIES have been outsourcing cementing since the 1920s, when Halliburton employees would appear at wooden derricks in trucks loaded with cement, a mixer, and pumps. BP adopted the outsourcing model with a vengeance, in line with its corporate doctrine of "decide which projects to back, and where possible have others do the work."[1] In contrast to companies like Chevron, which maintains its own cement-testing laboratory,[2] at the time of the blowout BP had just two corporate cementing sector specialists on staff—one for the Western Hemisphere and the other for the rest of the world.

Outsourcing creates tension between an engineering company's financial and technical arms. Outsourcing reduces costs, because the company does not need to maintain facilities and personnel: it simply pays for services as needed. However, outsourcing weakens the company's technology base—possibly to the point of nonexistence, which means that it does not have the expertise to properly monitor and control subcontractors. Subcontractors may also be reluctant to inject safety-related signals into the management control system, because such signals often convey bad news and may jeopardize the subcontractor's position.

BP exemplified this tension. As with its drilling practices, BP produced comprehensive standards for cementing but devoted little or no resources to enforcing them, among either its own employees or subcontractors. The latter included Halliburton, which provided a technical representative to BP Town and cementing technicians on the rig, backed up by a small laboratory in Lafayette, Louisiana. Halliburton also provided OptiCem, the proprietary computer program used by the company's technical representative.

The lack of oversight by BP played out in a weak management-of-change process used to vet Halliburton's multiple plans to cement the well—worsened by the shortcomings of the OptiCem system and the hesitant performance of the Halliburton representative.

OPTICEM AND ITS USER

After the blowout, BP asked an independent laboratory to evaluate OptiCem as part of an overall analysis of cementing on the *Horizon*. The laboratory raised serious doubts regarding the program's complexity, sensitivity, and sophistication: "It incorporates a number of complex engineering relationships that are not known or explained to the user. The complexity of the software is exaggerated with the addition of foam cement to the design. This complex fluid with properties that vary with pressure complicates calculation of density and friction pressure. In reality, friction pressure and density are coupled in a way that may not be fully described by the software. Further issues of eccentricity and erodibility are good engineering design components, but they may not mimic exact job performance. Discrepancies between design and actual conditions may be a result of engineering assumptions incorporated into the software rather than a job performance issue."[3]

BP's Western Hemisphere sector specialist similarly viewed OptiCem's approach as somewhat mindless, as indicated by his reaction to its cementing plan for a different well: "It seems to contain a cost recommendation for the recommended shiny designs and volumes, a lot of pages of Opticem [*sic*] output and some temperature modeling results. To me this provides an interesting example of a [design document], lots of canned output generated with the 'generate report' button, no discussion of job objectives, isolation objectives, design constraints/boundary conditions, why what was recommended was recommended, why technology was or was not applied, what the technology is that was applied, etc."[4]

BP's other cementing sector specialist expressed doubt to his Western Hemisphere counterpart about Halliburton's representatives themselves: "Welcome to my world, [the representatives] are typical of Halliburton engineers, they work well with BP Drilling engineers (they know how to keep them happy and don't make their life difficult)."[5]

In fact, BP Town was not all that happy with its representative for Macondo, as indicated by email from the junior drilling engineer while the representative was helping to plan the cementing of the production casing at the bottom of the well: "We aren't getting quality work from him anymore and I feel we need to push for someone qualified to take his place sooner rather than later."[6]

The next day the junior drilling engineer expressed frustration to the senior drilling engineer about the timeliness of the representative's work:

> I'm about to send this to [upper management], but wanted to send it past you first to make sure I'm not being out of line. [The representative] isn't cutting it anymore . . .
>
> I asked for these lab tests to be completed multiple times early last week and [the representative] still waited until the last minute as he has done throughout this well. This doesn't give us enough time to tweak the slurry to meet our needs . . . As a team we requested that he run another test with 9 gals on Wednesday, I know the first test had issues, but I do not understand what took so long to get it underway and why a new one wasn't put on right away. There is no excuse for this as the cement and chemicals we are running has been on location for weeks.

The senior drilling engineer responded, "Seems reasonable but a bit too late. We need to get his boss in and demand why his permanent replacement is not here."[7]

It was indeed too late. Three days after this email, Macondo blew out.

EVOLUTION OF THE PLAN

The cementing plan evolved from interactions between the junior drilling engineer and the Halliburton representative, with little input from others in BP Town, except for the misadventure involving the centralizers, and a management-of-change (MOC) meeting seven days before the blowout.

The first such interaction occurred at the end of March, just as the Macondo team recognized that the production casing would have to be 7 inches in diameter. The junior drilling engineer asked for a laboratory test of the proposed nitrogen-foamed cement mixture that the crew would use to attach the casing to the well, and the Halliburton representative produced a draft cement plan based on the estimated characteristics of the well. The representative then produced several more iterations of this plan—the last around midnight on April 18, distributed to BP Town and the rig crew just fourteen hours before cementing was to begin.

After submitting comments on BP's Kaskida petition for relief to the company's liaison to MMS, the operations manager took the step, highly unusual for BP Town, of scheduling an offsite meeting on April 14 and 15 to create an official MOC document for the production tail. Various individuals from BP Town attended, including the junior and senior drilling engineers, two levels of management from the engineering side, and the well team leader from the operations side, although that individual was out sick the second day. BP's cementing sector specialist attended for a few hours the first day, along with the Halliburton representative, who brought his laptop with the OptiCem software.

Oddly, the operations manager led what was in effect a pure engineering meeting, and made key decisions during it. That approach was an artifact of BP Town's awkward split between operations and engineering, as well as its practice of giving the operations group the lead once drilling had begun.[8] BP also outsourced the administration of the MOC meeting and the preparation of the official forms to a consulting firm, as was its practice—one that suggested to participants that MOC was something imposed from the outside.

A FLAWED REVIEW

The meeting was organized around a briefing by the senior drilling engineer.[9] The first two topics concerned the relationship between the well's pore pressure and the fragility of its formations. The third concerned whether to set a long production casing for a "keeper" well. The senior drilling engineer's slides gave a pessimistic view of the long-string option, and recommended the use of a liner—a shorter casing hung from an intermediate external casing string. The senior drilling engineer also proposed a third option: plugging the open hole with cement and abandoning the well without installing a production casing, leaving that for a completion rig. That "pure" abandonment approach, he said, would minimize the cost of the production tail activity— although the overall cost of the well would be higher—and provide more and safer options for completing it.

Although participants in the National Academies study of the blowout did not know about this proposal, they also noted that plugging and abandoning the well would have been safer: "When personnel on the rig encountered a low margin of safety between the ECD[10] and the fracture pressure, the safest approach would have been to plug the bottom open portion of the well and use the geologic data to design a replacement well. The replacement could have been a new well entirely or a sidetrack out of the lower portion of the existing well."[11]

However, the operations manager and other attendees rejected the senior drilling engineer's proposal.[12] Still, a revised version of his briefing prob ably used to brief management at some later time—again raised the topic of a simple abandonment: "Long string of 9-⅞" × 7" casing is *again* the primary option."

In a possible attempt to change management's mind, a second slide described the possibility of using a liner as a "contingency option," and a third stated that the plug-and-abandon approach was "*least preferred* (but still an option if hole conditions go south)."

This option included an estimate that plugging the open hole and returning later would add $10 million to $15 million to the overall cost of the well.[13]

In interviews with BP investigators two weeks after the explosion, the senior drilling engineer reiterated that the team could have chosen to temporarily abandon the well:

> [Investigator] then asked about the option to plug and abandon the well. [Senior drilling engineer] said that option was available. He said that he tried to convince [the operations manager] to consider it, as the team had met its primary objective for the Macondo project and was already behind schedule. He wasn't able to sell [the operations manager] on the concept.

[Senior drilling engineer] lone voice on just P&A[14] well
Could and should have run cement plug
The culture trained in 1989 did not support that
Culture—save money now[15]

During BP's investigation, the well team leader said he had supported the decision to continue with the production tail: "[Investigator] asked whether the team had seriously considered the [plugging and waiting] option, in light of the risks inherent in drilling this well. [The well team leader] said it was one of the possibilities considered but [he] personally believed it would have simply resulted in a deferral of the problem."[16]

The operations manager who had initiated the MOC team meeting expressed a similar sentiment: "[Investigator] asked if temporary abandonment (T/A) was discussed by the team. [Operations manager] said there was not a lot of conversation about TA—it wasn't preferred because if the well was TA'd, the next drilling team would need to deal with all the same risks, so if that if the team could get comfortable with the plan, it felt it should go ahead with the plan to complete the well. So, while TA was an option, it was not seriously discussed at the time."[17]

But events showed that deferral—time to review data from the well, conduct a risk analysis, and proceed based on that analysis—was precisely what the project needed.[18]

GO FEVER TAKES HOLD

There is no evidence that the senior drilling engineer appealed the MOC team's decision not to do a pure abandonment. He may not have thought his chances of success were very good. To make such an appeal, he would have had to go through two levels of engineering managers—each of whom had held his position for only eighteen days—and present his case to a vice president who had been on the job for just four months. And BP's incremental transition to a functional organization was not yet complete, so the company had no chief drilling engineer to whom the senior drilling engineer could appeal with some assurance that such a move would not adversely affect his career.

After the MOC team rejected the senior drilling engineer's proposal to plug the bottom of the well and return later, no one at BP ever reconsidered the decision to proceed with the production tail. Employees at BP Town debated only how to install the production casing, as evidenced by the decision tree the senior drilling engineer produced to document the options discussed at the MOC meeting (Figure 8.1).[19] (The decision process begins at the box at the center top, which defines a "conditioning run," a circulation to clean out the well bore.)

Because the MOC team had rejected the option of forgoing the production tail activity, all paths led to a well with a production casing—either a liner or a long string, either cemented successfully on the first try or fixed with a remedial "squeeze" cement job. One way or another, Macondo was to be finished, and the decision tree formalized that go fever.

In essence, the MOC exercise simply reaffirmed initial plans for constructing the well developed ten months earlier, which included the use of a long string and foamed cement to reduce stress on the fragile formation. The team did not consider an alternative to the latter because BP had used foamed cement successfully on four other wells and some team members thought it produced a better bond than solid cement.[20]

The main items that the team added to the initial plan were contingencies: a liner if the well's open hole collapsed during cementing, and a measurement called a *cement bond log* (CBL) if signals from the circulating system indicated problems while applying the foamed cement to the bottom of the production casing.

To perform a CBL, a crew of specialists lowers an instrument into a well and interprets readings to determine the presence or absence of cement.[21] The instrument cannot pass the float collar, so it cannot measure the cement below it, but it can indicate how high in the annulus the pumps had pushed the cement between the production casing and the formation.[22]

The MOC meeting was a contextual review, in that it considered the properties of the formation and cementing. However, it was incomplete in that it did not consider the number of centralizers and their placement. It also did not address any issues associated with the use of surplus casing material and fittings.

The junior drilling engineer left Houston for the rig on April 15, before the end of the MOC exercise. It is not clear why he was motivated to go to the rig that early, as the crew was engaged in the mundane process of circulating mud to clean out the well in preparation for running the production casing and cementing it.[23] He arrived just in time to walk into the middle of the centralizer distraction described in Chapter 7.

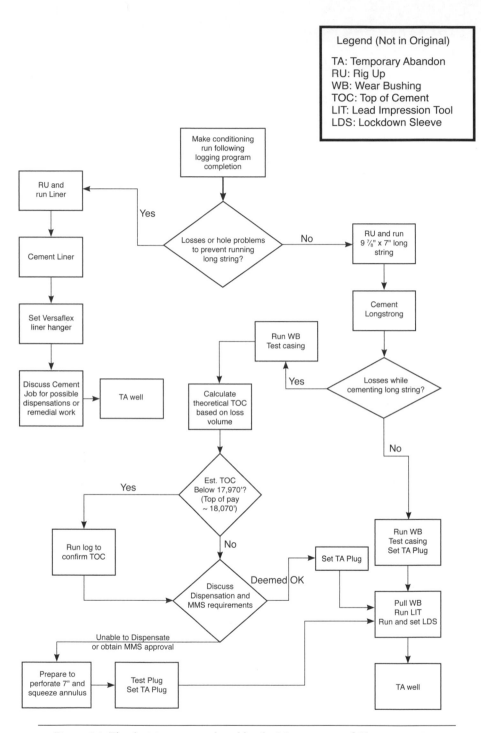

Figure 8.1. The decision tree produced by the Management of Change meeting.

A BRITTLE TOWN CRACKS

The next day, April 17, marks the point where it is no longer worthwhile to analyze BP Town as a control system, because the organization essentially ceased to exist as a functioning unit. Instead it had devolved into a collection of individuals who made decisions and issued directives with little or no coordination or even communication.

The first sign of this was the disorganized manner in which the matter of the additional centralizers unfolded. The second was this email, sent by the well team leader to his new boss (and old rival) that day:

> Over the past four days there has been so many last minute changes to the operation that the WSL's have finally come to their wits end. The quote is "flying by the seat of our pants". More over, we have made a special boat or helicopter run everyday. Everybody wants to do the right thing, but, this huge level of paranoia from engineering leadership is driving chaos . . . [The junior drilling engineer] has called me numerous times trying to make sense of all the insanity . . . This morning [the junior drilling engineer] called me and asked my advice about exploring opportunities both inside and outside of the company.
>
> What is my authority? With the separation of engineering and operations I do not know what I can and can't do. The operation is not going to succeed if we continue in this manner.

His boss made an attempt at placating the well team leader: "It should be obvious to all that we could not plan ahead for the well conditions we're seeing, so we have to accept some level of last minute changes. We've both [been?] in [the junior drilling engineer's] position before. The same goes for him. We need to remind him that this is a great learning opportunity, it will be over soon."[24]

And it was.

CHAPTER NINE

The Test and Displacement Plan

THE PLAN FOR TESTING the integrity of the cement that isolated the production casing from the formation—and then displacing the drilling mud in the upper well and riser with seawater, to prepare it for temporary abandonment—entailed complex and delicate maneuvers during which the crew risked losing control of the well. The plan for these tasks went through six revisions in seven days. None of these plans were subject to a management-of-change process, risk register, or any other form of contextual review.

The junior drilling engineer produced four plans in one format and a summary email, while the on-rig *mud engineer* from MI-Swaco, whose normal duties were to manage the drilling fluids provided by his employer, produced two one-page procedures—the last of which he briefed the drill crew on an hour before they were to begin executing it. The result was a set of procedures that not only had intrinsic dangers but were not fully accepted, and possibly not fully understood, by those who were to perform them.

A major spur to the revisions was the evolving response of the junior drilling engineer to pressure from BP's subsea organization to have a rig crew that specialized in exploratory drilling install the lockdown sleeve, used to secure the production casing.[1]

INFLUENCE OF THE LOCKDOWN SLEEVE

Although setting the lockdown sleeve was not part of the procedures for testing and displacement, the method chosen to secure the lockdown sleeve exerted a strong and adverse influence on the plan. The shape and function of the lockdown sleeve is shown in Figure 9.1.

The left-hand diagram shows the production casing and the casing seal set after the crew cements the bottom. The seal is left open during cementing, to allow the drilling mud displaced by the cement to flow from the annulus between the production and outer casings up into the riser. The seal is then

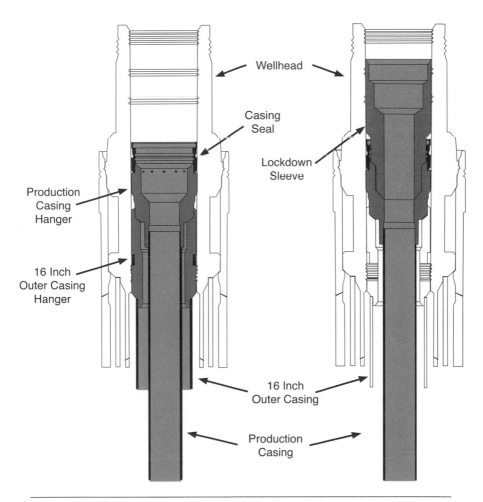

Figure 9.1. The lockdown sleeve.

closed to seal the annulus. At this point the casing is held in place by its weight and the cement at the bottom. The right-hand diagram shows the position and function of the lockdown sleeve, which is pressed into position to secure the production casing—a step that requires 100,000 pounds of downward pressure. BP chose to generate this pressure in a manner that adversely affected the later plan.

The first question associated with the lockdown sleeve is why the *Horizon* crew was setting it at all, as the crew of a completion rig normally performs that operation while hooking up a well to production facilities. BP Town originally assigned responsibility for setting the sleeve to the crew of the *Marianas,* presumably owing to its experience in drilling production wells. The Macondo risk register includes no entry for this task, and it devolved to the *Horizon* through the familiar BP process of agreements between individuals, with little or no contextual review.

When BP pulled the *Marianas* off Macondo in December, the head of the subsea organization asked if the *Horizon* crew would be setting the sleeve. The initial response from the senior drilling engineer was that the *Horizon* crew would run the casing but defer installation of the lockdown sleeve. The subsea manager responded by arguing that having the *Horizon* crew do both would save time, and therefore money: "Just FYI, we have installed a LDS from the *Horizon* on a well last year and everything went very smooth and took a minimal amount of time. I know you are aware but just want to reiterate that doing it when you drill the well originally saves an incremental 5.5 days of rig time when we get the wellhead ready for completion activities."

After a bit of back-and-forth and discussion of financial considerations, the senior drilling engineer agreed to this plan.[2]

The design of the sleeve required BP Town to find some means of generating the 100,000 pounds of downward force needed to secure it in place. The method the junior drilling engineer chose was to hang roughly 3,000 feet of heavyweight drill pipe from the sleeve. This approach was evidently driven by further cost concerns, because the crew used drill pipe already on the rig and set up on the drill floor in vertical "stands," ready to be run into the well. The other alternative was to commission a boat to deliver other forms of weight, and incur the time penalty of setting up that weight on the drill floor.

There then followed a cascading set of decisions—each influencing its successor—that led to the final plan. The subsea organization wanted the crew to set the lockdown sleeve just before displacement.[3] This meant the "surface" plug had to be set about 3,000 feet below the wellhead, to allow space for the weighting drill pipe.

This constraint opened two possibilities: to set the "surface" plug in drilling mud and then displace the mud with seawater; or to displace the mud in the riser and 3,000 feet of the well first, and then set the plug in seawater. Inexplicably—and as it turned out disastrously—BP Town decided on the latter course.

The decision to displace before setting the upper plug was inexplicable because setting cement in drilling mud is the normal procedure in wells across the world. The Macondo crew had used that procedure to cement all the outer casings and the bottom plug. The decision was disastrous because elementary well-control calculations show that about three-fourths of the way through the displacement, the well would become underbalanced.[4] That is, the push from the seawater column and the remaining mud would not be enough to resist the shove of the pay zone, should the bottom cement fail—which, in one manner or another, it did.

The plan also entailed placing the well in an underbalanced state with just one mechanical barrier in place to prevent hydrocarbons from escaping from the formation. That approach violated the "two-barrier rule," which at the time BP left to rig crews to define and enforce.[5]

THE TWO-BARRIER RULE

Like so much of the tacit knowledge entailed in constructing an oil well, this rule arose from painful experience. Its spirit is consistent with the industry maxim "Never trust a well." In this case, the rule is "Never let a well go underbalanced unless you have at least two barriers between you and the hydrocarbons."

The rule was not uniformly understood by all workers in the industry. Some contended that a BOP with its annular preventer open provided one of the required barriers, because the crew could shut in the well at the first sign of a kick. Others contended that the BOP with annulars open did not provide a barrier, as the crew could miss signs of a kick.[6]

We base our analysis of decisions regarding Macondo on the conservative definition, because replacing drilling mud with lighter-density seawater is an inherently dangerous operation. In this case, full displacement of the 14.1-ppg drilling mud in the *Horizon*'s riser with 8.6-ppg seawater would have reduced push by roughly 1,500 psi.

BP's *Drilling and Well Operations Practice* manual stated the two-barrier requirement for this operation in cryptic terms: "Two temporary barriers are required for moveable or hydrocarbon bearing or overpressured permeable sections from surface/seabed."[7]

Transocean's *Well Control Handbook* was more explicit:

> After setting the initial casing string(s) or during workover operations, a minimum of two independent and tested barriers must be in place at all times.
>
> There shall be two independent mechanical barriers to flow . . . kill weight fluid may be used in place of one of the two mechanical barriers.[8]

Transocean received a reminder of the importance of this rule two days before Christmas 2009, when one of its rigs in the North Sea, *Sedco 711,* barely escaped a blowout while finishing up a well for Shell. The events leading up to this near-miss were, in retrospect, eerily similar to the sequence that led to the Macondo blowout.[9] A mechanical barrier at the bottom of the well was inadvertently opened, no second barrier was in place, and the crew was confused about what was going on until mud showed up on the rig floor. In this case the crew was able to shut in the well at the last moment.

Transocean management debated whether to issue an emergency advisory to all rigs in the company's fleet reminding crews of the two-barrier rule, or to treat the incident in a more routine manner. One email sent during this debate clearly expresses the intent of the two-barrier rule: "When displacing to a lighter fluid which will cause you to be underbalanced, the only way to 'know and monitor' the fluid column is to know exactly what you have in the hole (and where) at all times, which in turn would imply you should know when you go underbalanced. At this point the fluid column ceases to be a barrier and the only barriers you have left are the BOP and the mechanical barriers downhole—at that point, any flow = failed barrier."[10]

The company eventually issued an advisory on April 5 intended "to clarify the requirements for monitoring and maintaining at least two-barriers when displacing to an underbalanced fluid during completion operations."[11]

For unexplained reasons, the company did not distribute the advisory to its North American operations until April 15, and it did not reach the

Horizon crew in time to affect its plan for Macondo. In any case, crews did not have to fulfill the actions mandated by the advisory until June 2010.[12] No one will ever know if doing so would have made a difference in preventing the Macondo blowout.

DISCONNECTS BEGIN

Compounding the inherent dangers of the plan was the manner in which BP Town transmitted it to the rig. BP Town not only relied on disjointed directives, but also assumed that leaders on the *Horizon,* particularly the well site leaders (WSLs), were of the same caliber as those who had led the rig through its almost fifty prior wells. This turned out not to be the case. To understand why—and why the difference was important—we must review the joint Transocean-BP leadership structure on the rig.

The Transocean crew was divided into three departments: drilling, marine, and engineering. The drilling department, or "drilling side," conducted the actual operations on the well. The marine department kept the vessel hovering over the well. The engineering department performed maintenance on the equipment used by the other two. Supervising these three was a single *offshore installation manager* (OIM) who reported to a Transocean rig manager in Houston.

The tasks performed by the drilling side were handed down from BP Houston and overseen jointly by the Transocean OIM and one of the two BP WSLs on board and on duty at the time.[13] The relationship between WSLs and OIMs varies greatly in the industry, depending on contractual strictures and, most importantly, personalities. At one end of the spectrum are WSLs who act like clients of the contractor and set general direction while leaving details to the workers. At the other end are WSLs who act like foremen and define and supervise every small step. Some WSLs are somewhere in between. What is clear is that the relationship is not one of superior and subordinates, as explained by a senior WSL on the *Horizon:*

Q. In terms of what happened on the rig, did the buck stop with you, is what I'm asking.

A. What happens on the rig is—is a team effort. It's not just me. The buck doesn't—you usually have to agree on something.

Q. And what happens if you don't agree?

A. If we don't agree, then we'd call Houston and let them make—make the decision.[14]

This informal, consensus-based decision-making process had worked for previous wells, and it worked for Macondo until the end of the drilling phase. It worked largely because of the experience of two of the four WSLs assigned to the *Horizon* and the time they had spent on the rig.

A POOR CHOICE OF WSLS

BP ranked its WSLs based on safety, management of money, performance, teamwork, and leadership. Personnel judgments are inherently subjective, and it would be unwise to read too much into individuals' rankings. Still, the two senior WSLs placed near the top in BP's November 2009 ranking.[15] Both were later assigned to the rig that drilled one of the Macondo relief wells after the blowout[16]—a technical feat that involved hitting a 7-inch-diameter target at a distance of three and a half miles and a depth just shy of 18,000 feet.[17]

The rotation of WSLs on the *Horizon* was fourteen days on and fourteen days off, with a hitch change every seven days on a Wednesday. This ensured that a new WSL came on duty in the middle of the other WSL's hitch, so they had seven days of overlap before another change. One senior WSL or another was on the rig three weeks out of four, and during one of those weeks both were on board. They split the day and night tours. The newly arriving WSL became the night WSL, the night WSL moved to days, and the day WSL went home. This was an unusual system: in most of the industry, the day WSL is senior and the night WSL junior, and the latter has instructions to wake the day WSL if anything important needs attention.[18]

One of the many "might have been" aspects of the disaster is that neither of the senior WSLs was on the *Horizon* at the time of the blowout. One was between hitches, and the other had cut his hitch short by four days to attend well-control school—something MMS required of WSLs every two years. Investigators and attorneys pressed BP management after the fact to explain why the company had not requested a waiver from the MMS to allow a senior WSL to remain on board the *Horizon*. BP managers responded that, despite its formal ranking system, the company regarded all WSLs as equally qualified and therefore interchangeable.[19]

BP knew two months in advance that one of the senior WSLs had to attend school and also wanted to attend his daughter's college graduation.[20] Despite the long lead time, the company chose the substitute WSL for the final four days of the senior WSL's hitch at the last minute. The substitute came from the *Thunder Horse*,[21] a large production rig that consolidated output from multiple wells and had started a maintenance break on April 9. The substitute WSL's only experience on the *Horizon* had been a visit eight or nine years earlier. He had only a few days to prepare before joining the *Horizon* crew, and he admitted that he "didn't really know the history of the well."[22]

The substitute assumed the position of day-tour WSL on April 16, arriving on the same helicopter that was to take the senior WSL he was replacing back to shore. The regularly scheduled night tour WSL had arrived the night before and was the most junior of the regular WSLs in time spent on the *Horizon*. This improvised shuffle broke the continuity of the normal rotation. Instead of having the senior WSL overlap for a week with the WSL rotating in, and having that person overlap for a week with the senior WSL's replacement, BP had two WSLs essentially rotate in simultaneously.[23] The result was that, from a pool of four regular WSLs and one substitute, the *Horizon* would be led by the WSL with the least experience on the rig and another who essentially had no experience with that rig and crew.[24]

There were personality issues as well. The junior WSL was described by one of his colleagues as "very competent"; the Transocean senior toolpusher, who directly supervised the drill crews, said of the junior WSL that he was "not a 'people person' and would not share information with anyone . . . He left things up to the rig to figure out."[25]

The supervisor of the substitute WSL had previously criticized him in an annual performance review when he was on *Thunder Horse:* "It sometimes appears that [the substitute WSL] is trying too much to impress the Houston office by attempting to have the answers to any questions that arise. There have been times when this has clearly not been the case."[26]

Town—particularly the well team leader—communicated with the WSLs primarily by telephone, starting with a 7:30 a.m. conference call called the "rig call," and following up multiple times during the day, depending on circumstances. This arrangement promoted disaggregation of information, as a WSL would rarely sit in on any contextual review Town might conduct of plans for the rig.

EVOLUTION OF THE PLAN

The process of planning the displacement and test portion of the production tail began on Monday, April 12, when the junior drilling engineer received an email from the senior WSL working the day tour: "We need procedures for running casing, cementing, and T&A work, we are in the dark and are nearing the end of logging operations."[27]

This email revealed the asymmetry of communication between Town and rig: even though the WSL was a BP employee and submitted daily drilling reports to Town, employees at Town did not keep him or the Transocean crew appraised of what they were doing or, more importantly, what risks they had decided the *Horizon* would assume. As we have seen, key personnel had made significant decisions in areas such as the lockdown sleeve as early as February but had not communicated them to the rig.

Even more significant was BP Town's reliance on the senior WSLs to interpret often cryptic instructions. That approach is especially significant regarding the "negative test," which, as we saw in Chapter 6, had no written definition. When the senior WSLs were not on the rig—as they were not on the day of the blowout—crew members had to recall what they had seen the senior WSLs do.

The *Marianas* Plan

The junior drilling engineer responded to the WSL's April 12 request for a plan with a draft by noon that day. The plan does not mention either the positive pressure test required by MMS or a negative test, and calls for installing the lockdown sleeve after setting a surface plug at the mud line. The document shows signs of being hastily assembled from some existing body of text, probably the original drilling plan prepared for the *Marianas*.[28]

The WSL responded with a reminder that the document did not include a negative test. He stated: "We need to do a negative test before displacing 14# mud to seawater."[29]

The significant phrase is "before displacing," indicating that the accepted practice on the *Horizon* was to test the well's ability to resist the pore pressure with the riser and entire well full of mud—the lowest-risk way of proceeding.

The junior drilling engineer promised a correction. He was also occupied on April 12 with producing material to support BP's petition to MMS for permission to delay the start of Kaskida,[30] and on April 13 he was concerned with the equipment to be installed in Macondo when the well entered into full production.[31]

Plan 1: Set Plug in Mud, Then Test

In the mid-afternoon of April 14 the junior drilling engineer sent an email to the rig with a one-page summary plan and asked the crew to incorporate it into the five-day rolling forecast of tasks to be performed. This plan— which presumably reflected decisions from the MOC meeting described in Chapter 8—called for setting the cement plug in mud at 8,367 feet and then performing a negative test to the wellhead depth using "base oil" in the kill line. The fact that the junior drilling engineer specified base oil in the kill line instead of seawater in the drill pipe suggests an unfamiliarity with the way the rig performed its negative tests.

Similar to diesel fuel, base oil is a fluid that is lighter than seawater. A column of base oil in the kill line would provide less push than a column of seawater, and would therefore allow more pore pressure shove on both the bottom cement job and the plug, providing a stronger test.

The senior WSL on duty reminded the junior drilling engineer that he had omitted the positive pressure test required by MMS, reinforcing the impression that the latter was confused about what had to be done.[32] Still, this was the safest of all the plans, because the crew would install a second mechanical barrier—the cement plug—before removing the drilling mud and its associated push from the riser and the well above the plug.

Plan 2: Test, Displace, Monitor

On April 15 the junior drilling engineer, newly arrived on the rig, sent a mid-afternoon email to the senior drilling engineer and others in Town with the question: "Recommendation out here is to displace to seawater at 8300' and then set the cement plug. Does anyone have issues with that?"

We have no indication who the individuals "out here" on the rig were who made this recommendation. In any case, late that night the senior drilling engineer responded: "Seems ok to me. I really don't think mms will approve deep surface plug. We'll see."[33]

And thus was initiated one of the most significant decisions made in the project: to displace the mud in the riser and upper well to seawater before setting the upper cement plug as the second barrier.

Several hours later the junior drilling engineer distributed an updated version of the twenty-one-page plan he had sent on April 12. The new plan reflected the MOC decision tree noted in Chapter 8: it called for a cement bond log to measure the height of the cement only if signals from the circulating system regarding pressure, flow, and volume during cementing clearly indicated failure. If they indicated anything else, the sequence after cementing the production casing would include performing a negative test with base oil, displacing with seawater from 8,367 feet, and monitoring the well for thirty minutes before setting the cement plug.

The wording of the steps is devoid of detail and subject to interpretation—so much so that when the chief counsel's staff showed identical wording to three BP supervisors after the blowout, they all disagreed as to what it meant.[34] The plan does not specify how much mud seawater would displace, and the phrase "monitor well for 30 minutes" could refer to performing a simple flow test with the annular preventers open, checking for mud being pushed out of the well using a sensitive volume measure called the "trip tank," or doing a full-blown negative test with the annular preventers closed. That is, the phrase could specify anything from a careful procedure to a violation of the two-barrier rule.

This plan was also shared with Transocean personnel on the rig, most likely the offshore installation manager.[35] Up to the last crew meeting on April 20, the OIM and other members of the Transocean drilling side likely believed that this plan was the one approved by MMS, instead of the one the agency actually approved—the next variation.

An Unusual Spacer

Meanwhile, senior members of the *Horizon* crew added yet another element to the already complex plan. Just before the end of exploratory drilling, the *Horizon* crew had mixed 450 barrels of two components of loss control material (LCM), in anticipation of using it, at which time they would add a third component. These two components, organic compounds with a water base, were deteriorating to the point that they could not be used at a later well. Environmental regulations required that they be taken off the *Horizon* and disposed of on shore. That meant a separate boat would have to come to the *Horizon,* because the *Bankston* would be full of used drilling mud, which also had to be taken ashore for recycling.

At some previous well, probably Kodiak, the MI-Swaco mud engineer and one of the senior WSLs had had a casual conversation about just such a situation. They had come up with the idea of avoiding the cost of an extra boat trip by finessing environmental regulations.[36] No one reviewed this idea in the context of the testing and displacement plan, either during its conception or later, despite the fact that the approach had never been done before, and no one knew whether this material would confuse signals from the circulating system during the delicate and dangerous displacement.

The regulations prohibited well crews from discharging LCM components into the Gulf unless they have been used: that is, circulated through the well. The mud engineer's and senior WSL's idea was to use the excess LCM components as spacer, to separate seawater from mud during displacement, so they could legally discharge it overboard. No one after the blowout appears to have asked why regulations include this loophole.

On April 15 this idea was incorporated into the plan. Not only had this material never been used this way before—either on the *Horizon* or anywhere else—but the 450 barrels was about double the amount normally used as spacer. Without the third element, the material was not supposed to set up, or polymerize, but no one performed laboratory tests or made calculations to verify this, or to assess the effect of the mixture on equipment and sensors.[37]

The MI-Swaco mud engineer mixed the two LCM components together, and added a thickening agent and enough barite to yield a density of 16 ppg.[38] The thickening agent affected the negative pressure test in ways that are in dispute. Some reports have argued that it could have confounded the test,[39] whereas others contend that although viscous, the mixture would have had no effect on the test.[40] However, notes for a briefing conducted as part of BP's internal investigation criticize the decision:[41]

> This was a deviation from procedure for which there is no documented MoC, and no evidence that any risk assessment was done for substituting Lost Circulation Materials for spacers in any combination. In reality, this could be

construed as "sham recycling," as a material was pumped downhole for a purpose it was not designed in order to change its status to exempt waste.

There was no apparent precedent for this action.[42]

Whether the material had any effect on the negative test will not be known until—and if—analysts comprehensively test the material under actual well conditions of pressure and temperature.

Plan 3: Test, Displace, Monitor

On April 16 the remaining senior WSL left the *Horizon* for well-control school on the same helicopter that had just delivered his substitute to the rig. The handover between the two men consisted of one phone call, two emails, and a thirty-minute meeting on the rig's helipad—clearly not enough time for the new WSL to learn the unwritten procedure for performing the negative test.[43]

Back in Town the senior drilling engineer documented the results of the MOC meeting in electronic form, attaching the decision tree and a design document.[44] He also produced an Application for Permit to Modify and sent it to the BP regulatory liaison for submission to MMS. The application included the following "forward plan," whose concise wording obscured its implications:

1. Negative test casing to seawater gradient equivalent for 30 min. with kill line.
2. TIH [trip in hole] with a 3-½" stinger to 8367'.
3. Displace to seawater. Monitor well for 30 min.
4. Set a 300' cement plug (125 cu ft. of Class H cement) from 8367' to 8067'.

   The requested surface plug depth deviation is for minimizing the chance for damaging the LDS sealing area, for future completion operations.

   This is a Temporary Abandonment only.

   The cement plug length has been extended to compensate for added setting depth.
5. POOH. [Pull Out Of Hole]
6. Set 9-⅞" LDS (Lock Down Sleeve)

7. Clean and pull riser.
8. Install TA cap on wellhead and inject wellhead preservation fluid (corrosion inhibitor) below TA cap.[45]

Because the crew had to set the cement plug in a uniform fluid, "Displace to seawater" had no possible meaning other than "Displace entire riser and upper part of well to seawater, going underbalanced in the process while relying on the single cement barrier in the bottom to prevent a blowout." There was also the ambiguity of "Monitor well for 30 min." noted before. Despite this, MMS did not request clarification, and approved the plan in less than a day.[46]

The senior drilling engineer then left work early for a weekend vacation, leaving the junior drilling engineer on his own out on the *Horizon*.[47] In the former's absence, multiple individuals would modify and interpret Plan 3 through a series of rapid revisions that would obscure the fact that those plans clearly violated the spirit of the two-barrier rule.

Plan 4: Displace, Then Test

The next day, Sunday, April 18, the junior drilling engineer sent an email to a number of individuals in Town asking if MMS regulations included anything on a negative test—another indication of his unfamiliarity with the process he had undertaken to devise. There is no evidence that he asked anyone on the rig, such as the WSLs or senior Transocean employees, for this information. He did not receive satisfactory replies to his emails, although, ironically, one quoted BP's two-barrier requirement.[48]

Later that day he emailed the well team leader with a summary of the plan as he understood it: "Plan is to do a negative test with base oil on the bottom plug. Then we will displace (a second negative test to greater value will happen) and following that set the cement plug."[49]

After which he asked if the crew could dispense with the first negative test. This email indicates that neither man was aware that Plan 3, submitted to MMS by the senior drilling engineer, had the first negative test conducted with seawater instead of base oil. The parenthetical remark suggests that the junior drilling engineer, at least, interpreted the cryptic "monitor well for 30 minutes" in both that plan and his as being a full negative test with the annular preventers closed.

Ten minutes later the well team leader responded: "I would use the sea-water displacement as the negative test, as you stated, shut it down at the end and do a flow test."[50]

When questioned by a BP attorney during MDL 2179, the well team leader stated that he meant that the crew should partially displace, do a full negative test, then open the annular preventers and finish displacing until all of the spacer and mud in the riser had been replaced with seawater. This interpretation still fails to justify the procedure he and the junior drilling engineer agreed upon. Even if the well passed that negative test, they would be displacing the riser and going underbalanced with just one tested me-chanical barrier, in violation of the spirit of the two-barrier rule.

There were now two displacement plans. Plan 3, which called for doing negative tests before and after displacement, was the one accepted by the senior drilling engineer, the Transocean crew, the managers who had signed off on the MOC, and the regulatory authorities. Plan 4, which eliminated the negative test before the crew displaced seawater in the well to 8,367 feet, was the one the junior drilling engineer and the well team leader had pri-vately agreed on without informing anyone else. This confusion would cause misunderstanding and friction between BP and Transocean, and within BP, and obscure the dangerous nature of the plan. And there was more confu-sion to come.

Plan 5: Details from the Mud Engineer

BP had given the onboard mud engineer from MI-Swaco responsibility for defining the exact procedure for displacing some of the heavy drilling mud in the riser with lighter seawater, although he later admitted to having only a general understanding of how to maintain control of a well during that risky procedure.[51] His main task was to calculate the volumes of sections of the well and riser, and convert those into the number of pump strokes re-quired to fill those sections in the specified manner.[52]

The mud engineer testified that creating such plans was part of the ser-vices he and his firm performed for clients such as BP.[53] Industry practices vary widely with regard to delegating such a sensitive task to a single indi-vidual from a services company. A small, single-rig operator working on land might be likely to assign such a task to an outside contractor. But for a deep-water, high-temperature, high-pressure well—especially one as difficult as

Macondo—a drilling engineer in Town would more prudently define this safety-critical task. That definition would include calculations predicting pressures, flows, and changes in pit volume during various stages of the procedure.[54]

Early Monday morning the mud engineer sent an email to his superior on shore with an attached one-page displacement procedure that he had produced by filling out a form on his computer.[55] The plan used the wrong capacity for the drill pipe and the overall well, and omitted the negative test. These errors may have occurred because he modified an existing document. Because his procedure omitted the negative test, it is impossible to tell whether he was working off the "official" Plan 3 or the "unofficial" Plan 4—although the latter is more likely because both he and the junior drilling engineer were on the rig.

Whichever plan he was using, he refined it in several ways. He specified that the spacer was to be displaced with seawater 500 feet above the BOP, and provided the number of pump strokes required to do that. He had the crew stopping the pumps just as the LCM spacer reached the rig. His plan also called for performing a static sheen test to determine whether the spacer was contaminated with oil based mud before the crew lined up the piping to dump it overboard. There is no evidence as to why he included this test, as it was not required by any regulation. The specified procedure concluded with "Mud engineer will advise"—evidently meaning that he would determine when displacement was complete, an uncommon delegation of a decision that the WSL would normally make.

More Confusion

At some point before the 11:30 a.m. meeting of crew members whose twelve-hour tour started at noon that day, the mud engineer must have given this version of his procedure to the substitute WSL, who evidently did not recognize the significance of the missing negative test. The Transocean senior toolpusher said that the substitute WSL was a "competent leader" but that he was "not sure about his knowledge of our operations and how we did the negative test, the displacement, and the pumping of the spacer."[56]

Evidence indicates that the substitute WSL must have reviewed the mud engineer's plan during the meeting, which prompted a minor confrontation with the Transocean offshore installation manager, who flatly refused to displace mud out of the riser without a negative test. The substitute WSL conceded that the plan should include the test.[57] The OIM likely left the meeting with the impression that the negative test would be added to the mud engineer's procedure where it was in Plan 3, and where the crew had performed it during all the displacements it had done to date: before removing significant amounts of mud and thereby reducing the amount of reserve push in the riser.

The evidence suggests that sometime on the morning of April 20—possibly during the 7:30 a.m. "rig call" conference call—the senior drilling engineer realized that the displacement sequence of "test then partially displace then test again" that he had devised and arranged for MMS approval was not the one the substitute WSL was describing to him. Instead, the substitute WSL indicated a "partial displace then test" sequence—Plan 4, the one that the junior drilling engineer and the well team leader had agreed upon on Sunday, and one that the senior drilling engineer had never heard of. He called the substitute WSL and told him that.[58]

The substitute WSL went and woke up the junior drilling engineer and asked him about the discrepancy. The junior drilling engineer told him that "the team" had agreed to eliminate the negative test before displacement while the senior drilling engineer had been away on vacation. This sequence was further evidence of the communication gap between Town and rig.

Meanwhile, back in Town the senior drilling engineer was meeting with the well team leader and two engineering managers to discuss the plan. The outcome of this meeting was that Plan 4—the procedure approved by the well team leader that eliminated the negative test prior to displacement—would replace Plan 3, the longer one approved by MMS. The senior drilling engineer later told an interviewer that "[the well team leader] was hard to argue with." The group also decided, for reasons not reflected in the evidence, that BP did not have to inform MMS of the change. The senior drilling engineer then called the rig and informed the substitute WSL of the decision.[59]

The substitute WSL then met with the mud engineer and verified the procedures they were to follow.[60] By midmorning on the day of the blowout, the BP and MI-Swaco individuals involved with the procedure were finally on the same page. The junior drilling engineer wrote it up as a one-page "Ops note" in an email, which he sent to all interested parties within BP just before he left the rig on the 11:30 a.m. flight.[61] It is clear from the evidence that no one from BP had bothered to inform any Transocean employees—who were responsible for the actual testing and displacement—of the new procedure.

At the noon meeting that day, the day driller and the OIM briefed the crew on the displacement procedures to be followed that afternoon, which they still believed to be the "test then partially displace then test again" sequence of Plan 3. They were interrupted by the substitute WSL, who told the assembled crew members "Well, I just got a different procedure."

He then outlined Plan 4's "partially displace first and then test" sequence agreed upon within BP just hours before. The OIM objected, reportedly saying, "Well, we've never done something like this before. We've—I don't remember ever doing this before, and I don't think it's going to work."

And then, in a clear reference to Plan 3, the OIM added, "This is what I've got that's approved from MMS and until I see something different, this is what we're going to do."

The substitute WSL insisted on the new procedure, and some witnesses reported that the exchange grew heated, at which point the day tour driller cut the exchange short and said, "We'll work this out on the drill floor."

As the meeting broke up, witnesses testified that the OIM said, "Well, I guess that's what we've got those pincers for."

This was a clear reference to the possibility that the crew would have to activate the blind shear rams to cut the drill pipe and seal off the well in response to a kick.

There is no doubt that, once on the drill floor, the substitute WSL succeeded in imposing Plan 4 on the Transocean drill crew. The day subsea superintendent, who assisted the crew in the drill shack by operating the BOP, testified that the day toolpusher expressed dissatisfaction with deviating from the tried-and-true sequence of performing a negative test before displacing any drilling mud with the lighter seawater.[62]

At 1:09 p.m. the well team leader called the substitute WSL from BP Town.[63] There is no evidence or testimony as to the nature of the call, but a reasonable surmise is that he called to verify that the crew understood and accepted the new procedure, which entailed partially displacing before performing the negative test.

Plan 6: The Final Procedure

At 3:30 p.m. on April 20, some six and a half hours before the blowout, the mud engineer briefed the crew on the drill floor on Plan 6, an updated version of the one-page plan he had sent to his management on Monday, and which had triggered the first confrontation between the substitute WSL and the OIM. This plan is documented in the evidence as an attachment to an email sent to the substitute WSL just before the briefing.[64]

The plan calls out the negative test, but like all the plans does not provide detail on how the crew is to perform it. The volumes for the various elements of the well have been corrected, but the volume for the capacity of the drill pipe is less than that later calculated by forensic analysts. That error is significant because the number of pump strokes specified to raise the spacer above the BOP is barely sufficient based on the mud engineer's volume for the drill pipe. This, and later events, would place the viscous spacer in the kill line and below the BOP, possibly confounding the negative test.

The plan contains a second error: the number of pump strokes required to bring the spacer just barely up to the rig for the static sheen test is too large, and actually represents the number of strokes required to displace all of the spacer out of the riser—a number that correctly appears in an earlier paragraph on the page, showing the effect of probable haste and lack of review.

The convoluted process by which the integrity test and displacement plan was developed stressed the fragile command structure on the rig and obscured the fundamental flaw in the plan without yielding any benefit. The process diverted attention and resources away from more important concerns, such as the need to perform the predictive well-control calculations that would have exposed the risk inherent in the plan. Instead the displacement and test plan became an element in an overall temporary abandonment plan—one that, as we shall show in Chapter 10, piled risk upon risk until a trip over The Edge was all but inevitable.

CHAPTER TEN

A Reconstructed Plan

IN THIS CHAPTER we reconstruct a consolidated plan from the parallel sequences of disaggregated decisions described in Chapters 7, 8, and 9. For each step in the plan we explore the risks that BP Town did not explicitly analyze, alternatives they did not consider, and the results of predictive well-control calculations they did not make.[1] Those calculations—the most basic in the field—show the displacement to seawater in the plan to incur risks that would be unacceptable to most experienced drillers at the time, and which today are illegal in the Gulf.

This chapter is admittedly from a "luxuriant retrospective position," and unfair to those who, at the time, had to struggle with these issues in the context of a brittle organization operating under stringent time and cost constraints. That being said, we need to explicitly consider BP's tacit plan, to understand how far BP Town's way of operating was from the ideal, and to appreciate the well-control risks the plan posed for those on the *Horizon*—risks obscured by the fragmented and disordered manner in which Town communicated tasks to them.

The sections that follow lay out the plan in the way it could have been presented at a proper contextual review, starting with how the well was supposed to look after the planned activities were complete. We then examine the serious risks entailed in each step in that plan—risks that no one from BP appears to have evaluated at the time.

AFTER ABANDONMENT

The configuration shown in in Figure 10.1 is the one the Macondo project intended to leave for a rig that specialized in completions. That rig would have installed piping and controls to allow oil to flow along an underwater pipeline to a production facility several miles away, where it would have been brought to the surface and loaded into tankers for transport to whatever refinery was to process it.

This configuration was driven by the decision to install both a production casing and a lockdown sleeve. As we noted in Chapter 8, the team had the option—as proposed by the project's senior drilling engineer—to set plugs at the bottom and the top of the well, and either to abandon it and drill a new well nearby or to drill a sidetrack into the formation. This course of action would have reduced the risk of a blowout to acceptable levels. The well team leader and the operations manager rejected this option.

Another alternative would have been to allow the crew to set the upper plug before displacing the heavy mud in the riser and upper well with seawater. In that case, even if the cement at the bottom of the well was so poor as to leave the production casing fully open to the pore pressure, the mud remaining in the lower well would have produced enough push to limit pressure on the upper plug to about 1,000 psi, and even less on the seal where the production casing met the wellhead. Those parts of the well could have easily resisted both these pressures.

Neither of those pressures would have posed any problem to the crew of a completion rig, which would have installed its own BOP and riser and displaced the seawater in the well with mud. The completion crew would then almost certainly have performed a negative test, which would have alerted them to the need for remedial cementing at the bottom of the well. The completion crew would also have performed that procedure with a riser and well full of mud—and thus minimal danger of a blowout.

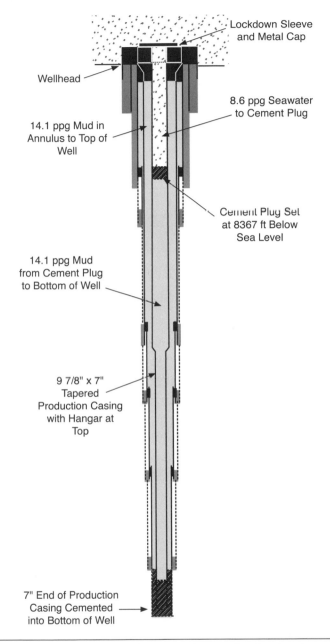

Figure 10.1. The planned configuration of Macondo after abandonment.

RUNNING CASING

As noted earlier, the original design for the production casing envisioned a 9⅞-inch "long string" running from the top to the bottom of the well, and BP had procured the needed casing material and fittings for the well. When the fragility of the Macondo formation required the use of two contingency strings, the diameter of the external casings shrunk to 9⅞ inches, prompting the hasty procurement of surplus casing and fittings. (See Figure 10.2.)

At this point the team had the option of running a "liner"—a short production casing attached to the external casing at some point deep in the well instead of at the very top. Participants in the same meeting where the operations manager and the well team leader rejected the idea of permanently abandoning the well considered and then rejected this option, on the grounds that a liner had a greater risk of failure after long periods of production.

The production casing was—according to standard practice—to be assembled in situ, one section at a time. First thirty- to hundred-foot "stands" of casing weighing upward of 3,000 pounds are each hoisted upright over the entry to the well and threaded on the stand that precedes them. The assembly is then partially lowered by the draw works and gripped by viselike slips so the process can be repeated for the next stand. When the last, topmost stand is in the slips, the casing hanger is attached. Then the casing is lowered in place, as described in Chapter 7.

During the running of casing, the well and riser are full of 14.1-ppg mud, and the crew faces no significant well-control issues. However, running casing does entail risk. Assessing the depth of a two-and-a-half-mile-deep hole is an inexact business unless the crew uses a wireline to perform that measurement.

At Macondo, a shelf of rock sticking into the hole at about 18,200 feet prevented the wireline tool from reaching the bottom of the well. As a result, the assumed last 160 feet or so of the well was not measured, and its exact depth was unknown. If the well was shallower than the drillers thought, the casing would be too long and strike the bottom of the well or residual cuttings, and buckle, opening a breach in the casing that would later allow hydrocarbons to flow into it and possibly cause a blowout. The difference between properly placing the casing and hitting bottom at Macondo was

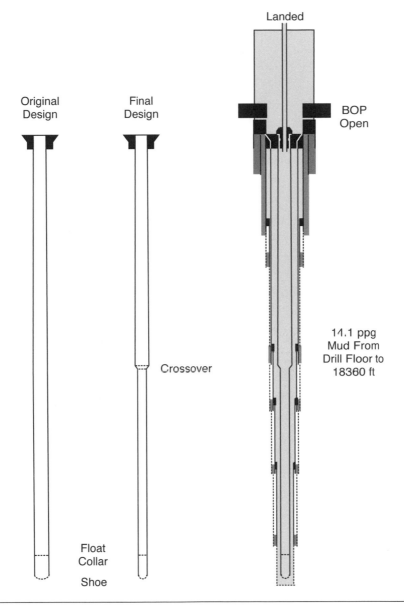

Figure 10.2. Original and final configuration of the production casing.

fifty-four feet in two and a half miles—well within the margin of error entailed in estimating the depth of the well.

CEMENTING

BP Town had chosen to use foamed cement from the very beginning of the project, because of fear that the fragile Macondo formation could not take the strain entailed in pumping heavy cement. No one reopened that decision during the course of the project. If the formation crumbled, the cement would flow into it just as mud does during lost returns, and there would not be enough left in the annulus between the production casing and the formation to provide a solid barrier against the flow of hydrocarbons. (See Figure 10.3.)

Rather than trace the convoluted history of the cementing plan, we will summarize the shortcomings of the cement job uncovered by multiple analysts after the blowout. These fall into three broad classes: inadequate testing of the cement mixture before it was used, the effects of dealing with the fragile formation, and the improper use of foamed cement.

Inadequate Testing

Analysts criticized BP for starting the cement job before laboratory tests verified the ability of the slurry to set up and its strength after it had set up. The setting of cement depends on temperature, and analysts criticized Halliburton for testing at unrealistic temperatures, which gave incorrect results. Two analyses concluded that, as a result, the cement had not set when the crew performed the negative test.[2]

Risky Procedures

The first effect caused by the nature of the formation was the use of nitrogen-foamed cement to reduce the density of the material from the normal 16-ppg range to something under the 14.5 ppg that the well could tolerate. The design for the bottom of the well also called for an unusually small amount of cement pumped at unusually low rates, which reduced the chances that it would properly flow up the annulus and seal off the hydrocarbon zones.

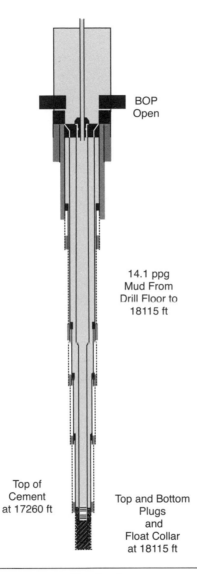

BOP
Open

14.1 ppg
Mud From
Drill Floor to
18115 ft

Top of
Cement
at 17260 ft

Top and Bottom
Plugs
and
Float Collar
at 18115 ft

Figure 10.3. The plan for cementing.

Finally, the *Horizon* team decided that it was too risky to fully circulate out the stagnant mud, cuttings, and other possible debris in the bottom of the well. The crew did a partial circulation instead, raising the risk that the cement would become contaminated as it was pushed down the shoe track and "turned the corner" to flow up into the annulus.

Improper Application of Foamed Cement

Through some undocumented process, BP Town decided to use dry cement left over from the Kodiak well as the basis for the cement slurry used to cement the production casing. This mixture contained an antifoaming chemical, which increased the risk that the cement would not expand into the space between the production casing and the formation.

Far more serious was the attempt to apply foamed cement in a well filled with oil-based mud. In the words of one of the expert witnesses: "The destabilizing effects on foamed cement by [oil-based mud] are severe and can lead to a job failure. The risks of failure are so severe that I have not, nor will I, recommend using foamed cement in an oil-based mud environment."[3]

"Destabilizing" in this context means that the nitrogen escapes from the foam, leaving a smaller residue of cement slurry, like soapsuds gradually collapsing into a film. The destabilizing effect of oil-based mud becomes more severe when used in a well such as Macondo, whose fragile formation was eroded or "washed out" by the circulation of mud during and after drilling.

The crew detects washout using a wireline measurement called a caliper run, in which feelers sense the diameter of the well as the measuring tool is pulled through it. Figure 10.4—adapted from the actual caliper run from Macondo[4]—captures the effect of the washout in one of the sections to be cemented. The dotted line shows the intended hole diameter of 9⅞ inches, and the uneven lines show the results of the caliper measure, with the production casing in the center. The effect of the jagged surface is to trap oil-based mud as the cement is pumped up past it. The already unstable cement mixture is then further destabilized, shrinks, and fails to seal the gap, while gelled mud prevents the cement from adhering to the formation.

The result is that there are two possible paths from the formation into the production casing: through a casing breach or down alongside the casing, into the shoe, and up through poorly set or contaminated cement inside the shoe track. The slow pump rate chosen to avoid fracturing the formation increases the probability that the washed-out sections will trap oil-based mud. No number of centralizers would have prevented these effects.

There is no evidence that BP Town considered any of these issues at the MOC meeting or—distracted by the centralizer issue—at any later time. The well team leader admitted to BP investigators that BP Town implicitly accepted the cementing approach: "[The well team leader] noted that [the

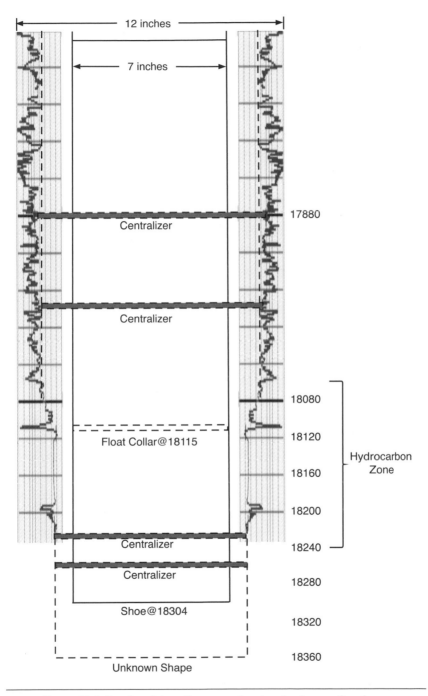

Figure 10.4. Results of the caliper run, showing the degree of washout.

Halliburton representative] was known for 'just in time delivery' of test results and that if [the representative] delivers the results later in the evening or overnight, [the well team leader] would not see those until the next morning when he gets to work. He confirmed that there were no objections raised to the cement program by anyone at the 7:30 am calls on April 17–18."[5]

Expert witnesses, BP's own cementing sector specialist, and others have observed that BP Town could have considered other approaches to dealing with the fragile formation, such as mixing the cement with loss control material (LCM) to keep the cement from flowing into the formation before it had set.[6] However, these options remain speculative. It is possible that a production casing could not have been cemented in Macondo using any known technology, and that BP's only option was to either sidetrack or permanently abandon and drill a new production well nearby.

Despite these risks, as with running the casing, the well and riser were full of 14.1-ppg mud during cementing, so the procedure would not have entailed well-control issues.

FIRST DISPLACEMENT

During this operation, the procedure chosen by the junior drilling engineer and the mud engineer had the crew pumping 16-ppg spacer into the well, followed by seawater, until the well assumed the configuration shown in Figure 10.5. The mud pushed out of the riser was to go back into the pits that stored mud on the rig. The rationale behind this task was to displace down to where the upper cement plug would be set in seawater, so the negative test to immediately follow would more accurately simulate the hydrostatics during that cementing.

The alternative to this and the next three steps would have been to set the upper cement plug in mud, and then displace with seawater and do the negative test. That approach would have entailed placing two barriers in the well before displacement—fulfilling a basic well-control requirement. The evidence provides no rationale for the decision to set the plug in seawater.

Basic hydrostatic calculations for this configuration indicate that the combined hydrostatic push of the spacer, seawater, and mud would have been

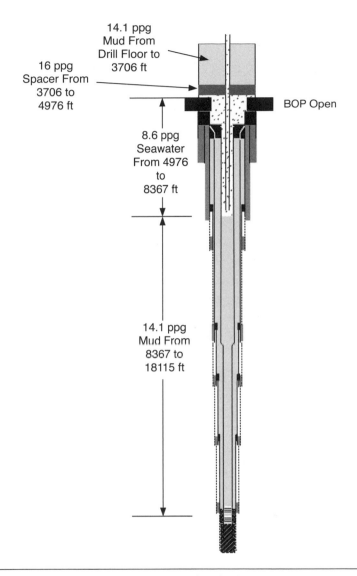

14.1 ppg
Mud From
Drill Floor to
3706 ft

16 ppg
Spacer From
3706 to
4976 ft

BOP Open

8.6 ppg
Seawater
From 4976
to
8367 ft

14.1 ppg
Mud From
8367 to
18115 ft

Figure 10.5. The planned first displacement.

enough to overcome the pore pressure, and there would have been no influx of hydrocarbons into the well—and thus no blowout. However, the combined push would also have been very close to the amount that would have fractured the formation and allowed mud to flow into it. If not detected, that would have led to a drop in hydrostatic push and possibly allowed an influx of hydrocarbons into the well—an outcome that may well have occurred during the actual procedure, after some change in the well that permitted a flow out of the production casing.

NEGATIVE TEST

Basic well-control calculations predict that a failed cement job that allowed hydrocarbons to flow into the production casing would produce a reading of slightly over 1,000 psi at the pressure gauge, and continuous flow when the valve was opened at the cement unit for the negative test—both clear indications of failure. (See Figure 10.6).

The *Horizon* crew traditionally ran this test by displacing mud in the drill pipe down to just below the BOP, as shown in Chapter 6. If the crew had run the test that way—replacing the mud in the drill pipe with base oil, as specified in Plan 1—basic well-control calculations show that the pressure gauge in the cement unit would have read almost 800 psi, and flow out at the cement unit would have been continuous. If the crew had used seawater in the drill pipe, as specified in Plan 3—the one approved by MMS—the pressure gauge would have shown a pressure of just over 200 psi, and flow would again have been continuous. These calculations show that any of the three options for conducting the first negative test—base oil to the wellhead, seawater to the wellhead, and seawater to 8,367 feet—would have shown that the cement barrier at the bottom of the well had failed.

As with the centralizer distraction, changing the negative test sequence to one that called for partial displacement first was wholly unnecessary. All the change of sequence achieved was confusion and contention within BP Town, and between BP Town and the Transocean crew on the *Horizon*. The change added to the disorder that prevailed on the drill floor at the time of the negative tests, and that disorder contributed to the fatal misinterpretation of signals from the circulating system during the tests.

FINAL DISPLACEMENT

During this task, the crew was to pump seawater into the well to provide a uniform fluid column for setting the upper cement plug (Figure 10.7). The returns out the top of the riser were to go back into the pits until the spacer reached the top, at which point the crew would stop the pumps, verify that

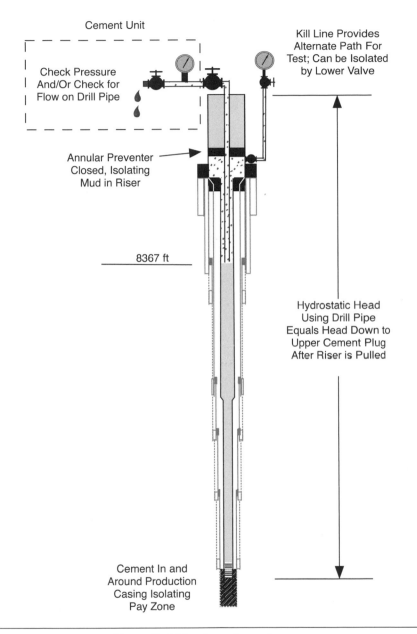

Cement Unit

Check Pressure
And/Or Check for
Flow on Drill Pipe

Kill Line Provides
Alternate Path For
Test; Can be Isolated
by Lower Valve

Annular Preventer
Closed, Isolating
Mud in Riser

8367 ft

Hydrostatic Head
Using Drill Pipe
Equals Head Down to
Upper Cement Plug
After Riser is Pulled

Cement In and
Around Production
Casing Isolating
Pay Zone

Figure 10.6. The planned negative test.

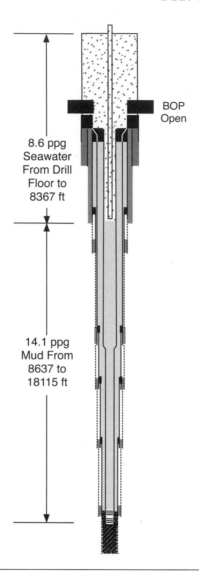

8.6 ppg
Seawater
From Drill
Floor to
8367 ft

BOP
Open

14.1 ppg
Mud From
8637 to
18115 ft

Figure 10.7. The planned final displacement.

the spacer was free of oil, and restart the pumps, discharging the returns overboard. This procedure was designed to exploit a loophole in environmental regulations that allows a rig to discharge unused LCM components overboard.

This displacement would occur with just one mechanical barrier at the bottom of the well—the cement, with the BOP as a potential second barrier.

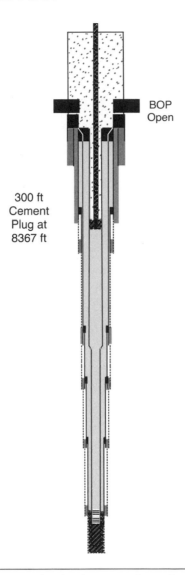

Figure 10.8. The plan for setting the surface plug.

However, because the BOP must be open for the displacement to occur, the configuration violates at least the conservative interpretation of the two-barrier rule.

Basic well-control calculations—again under the assumption that the production casing is open to the formation—predict that the well would become underbalanced when about two-thirds of the mud in the riser has

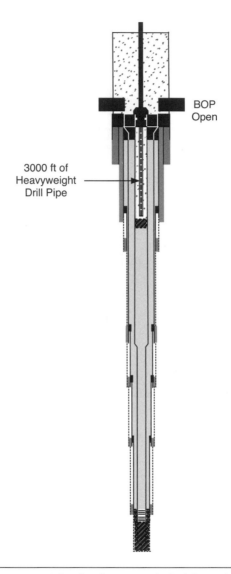

Figure 10.9. The plan for setting the lockdown sleeve.

been displaced. These calculations also show the roughly 1,000-psi imbalance between hydrostatic push and the shove of pore pressure that the negative test would have shown.

Again, there is no evidence that anyone in Town or on the *Horizon* performed this predictive calculation. That almost certainly means that no one on the project realized the danger posed by the combination of a single mechanical barrier at the bottom of the well, an open BOP at the top, and insufficient push from fluid in the well.

SETTING THE TOP PLUG

The crew never performed this step because Macondo blew out during the previous one. As we noted in Chapter 9, the inexplicable and unnecessary decision to set the top plug in seawater (Figure 10.8) led directly to the need to displace mud in the well and the riser with seawater with just one mechanical barrier at the bottom of the well—one that failed.

SETTING THE LOCKDOWN SLEEVE

The crew never started this operation either. The Chief Counsel's Report suggests that alternative ways to generate the weight required to set the lockdown sleeve would have been safer, because that would have required displacing 1,500 feet less mud in the well with seawater.[7] (See Figure 10.9.) However, basic well-control calculations show that displacing less mud would not have helped: the well would still have become underbalanced while the crew was displacing mud in the riser with seawater.

THE DISTANCE TO THE EDGE

Two significant decisions—to use foamed cement and to set the upper plug in seawater—had a cascading effect on other decisions. And that effect was obscured by rapid-fire changes during the planning process and BP Town's aversion to contextual reviews.

The final result was to move the *Horizon* and those aboard her significantly closer to The Edge. Responsibility for preventing the well from going over now moved to the command structure on the rig, supported by data from its circulating system. Both were ad hoc constructs that combined with time pressure, go fever, and the distractions of a VIP visit to produce the emergent event that was the Macondo blowout.

CHAPTER ELEVEN

The Systems of the *Horizon*

THE *DEEPWATER HORIZON* was designed in the late 1990s for a company called R&B Falcon, then the world's largest offshore drilling contractor. By the time of her launch in 2001, R&B Falcon had been acquired by Transocean (then called Transocean Sedco Forex), and the *Horizon* remained under the control of Transocean or a Transocean subsidiary until her sinking. During her lifetime she flew two different "flags of convenience"—countries of registry chosen for favorable tax and regulatory treatment: first Panama, and then the Marshall Islands.[1]

The design of the *Horizon* essentially duplicated that of an earlier R&B Falcon rig called the *Deepwater Nautilus*. The *Nautilus* was, and is, an anchored rig that must be towed from place to place. The *Horizon* design added equipment such as thrusters, which converted her into a self-propelled, dynamically positioned vessel.

The main element of the *Horizon* was her hull, a steel box slightly longer than a football field and about one and a half times as wide. In contrast to conventional vessels, the "hull" of the *Horizon* was never designed to be in the water. Rather, it rested on four large columns that were, in turn, attached to two large pontoons that never broke the surface of the water. The bridge crew could pump water in and out of the pontoons to alter the height of the rig above the surface—up for when she was moving, and down when the crew was drilling.

When the *Horizon* was latched up on a well, she was essentially a set of drilling equipment that happened to be afloat. When she was moving from well site to well site, she was a seagoing vessel that happened to be carrying a set of drilling equipment. The organization of her crew and equipment into the marine side and the drilling side reflected the two modes of operation.

The two sides had significantly different equipment and cultures. Together those elements formed two different control systems that were physically and operationally separate—a situation that complicated efforts to save the vessel after the blowout.

THE MARINE SIDE

Almost all employees on the marine side came from just one company—Transocean. They were led by a *master,* who was in charge of the bridge crew, and a *chief engineer,* who oversaw the power plant and maintenance of all equipment, including that used by the drilling side, supported by an onboard engineering team.

A single vendor, Kongsberg AG—acting as systems integrator—also provided all the technology used by the marine side. This control system operated out of two locations, the bridge and a lower-deck engine control room (Figure 11.1).

The bridge (Figure 11.2), also known as the *central control room,*[2] was a relatively spacious and modern facility. The bridge contained a bank of consoles for each of the major control functions: changing the ballast in the pontoons, controlling the engines, and maintaining station with the dynamic positioning system, as well as for gas and fire alarms. The bridge also had a panel for the BOP's emergency disconnect system. Closed-circuit television screens permitted observation of much of the vessel except the interior of the drill shack. The engine control room—located on a lower deck between two banks of three engines each, which kept the *Horizon* on station and provided electrical power for the drilling operations—backed up those critical control functions.

Federal law requires senior members of the marine side to hold licenses granted by recognized maritime states. The license held by a master requires

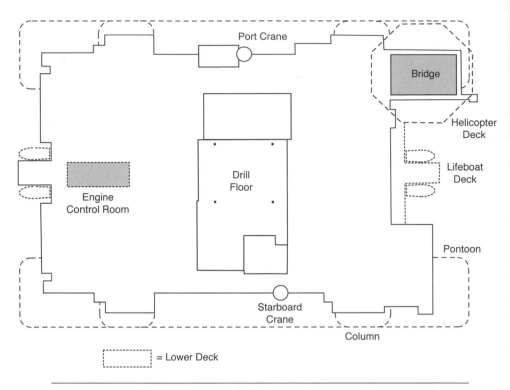

Figure 11.1. Layout of the main deck, showing locations of the marine side.

Figure 11.2. The interior of the bridge.

graduation from a four-year maritime academy and a minimum of three years spent working one's way up through the chain of command. The chief engineer's license requires five years of engine room experience.

The bridge crew operated according to the disciplines of the sea as defined by custom, law, and treaty. The crew had a stable set of duties and stood watch in a manner identical to the bridge crew of a ship crossing the ocean. Because the rig was stationary, boredom was an issue, and procedures required crew members to change physical positions on the bridge every hour.

Four subsea specialists on the rig tended to the BOP, the riser, and the diverter system, which routed fluids such as mud that exited the riser either overboard or into the pits where it was stored. Two were lead specialists, called senior subsea supervisors, and two were assistants, called simply subsea supervisors. Although part of the engineering group, these specialists also directly supported the drilling side.

The subsea specialists also performed maintenance on the BOP, based on Transocean's policy of "condition-based maintenance," which they called "condition-based monitoring." Both lead specialists defended that policy in their testimony. When asked about the chief mechanic's characterization of that policy in his testimony as "run it until it breaks," one specialist replied that such a practice was definitely not to be found in his department.[3] Instead, the subsea specialists argued, their practice of pulling the BOPs and inspecting them for maintenance issues every rig move was more effective than the manufacturer's recommended teardown at fixed intervals—this despite the evidence that their practice led to a dead battery in one of the two BOP control pods.

A final subsea activity was the use of miniature robot submarines called remote operated vehicles (ROVs) to swim down to the BOP or along the riser to perform inspections or minor maintenance. Oceaneering, a specialist contractor, owned and operated the ROVs, and the company's employees worked from a dedicated area on the forward starboard corner of the main deck. The *Horizon*'s ROVs were not deployed at blowout, but those from other vessels were active in attempts to stop the flow of oil afterward.

A TIRED RIG

Despite the contention of the subsea specialists, the engineering group struggled with a rig that was seriously overdue for major overhaul. The *Horizon*

had not been in drydock since its launch in 2001. In October 2009 the chief
engineer, in an impassioned email to the rig manager, referred to the *Horizon*
as "just limping along."[4]

Transocean's investigation after the explosion estimated that as of April 20
the *Horizon* had 1,900 employee-hours of maintenance tasks backlogged for
more than thirty days.[5] There is also no evidence that the sensors in the circulating system, such as those that measured pit volumes or pressure, had
been recalibrated at the five-year interval that regulations required. The log
of open maintenance issues noted that there was no evidence that pressure
gauges had been checked since the rig was put in service, and scheduled them
for calibration before September 2010.[6]

The maintenance situation was particularly severe for the electrical and
electronic systems on the drilling side of the rig, which had suffered from
years of quick fixes. Crew members arbitrarily rewired equipment to overcome a specific malfunction without later restoring the system to its original
condition. This exchange between a Halliburton lawyer and the chief electronics technician described the situation in 2010:

> Q. Let me ask you a question that, in retrospect, may seem like an odd
>   question, but up until the time of the incident, did you actually like
>   working on the *Horizon* rig?
> A. Yes, I did.
> Q. Okay. Why was that? Why did you like working on the rig?
> A. It was challenging.
> Q. Okay. How so?
> A. Every time I went to fix something, I would find layers of repairs and
>   Band-aids, and it was a—sort of a mystery novel to get to the bottom
>   and find the real underlying issues.[7]

The condition of the equipment on the *Horizon*—which a plaintiffs' expert
described as "deplorable"[8]—may not have led directly to the blowout. However, it added a layer of uncertainty to anomalous signals from the circulating
system that crew members were attempting to interpret: Is something going
on down in the well, or is the gear just acting up again?

During the final cement job, crew members ascribed anomalous pressure
readings to a malfunctioning pressure gauge, and there is no evidence that
anyone investigated further.[9] This lack of concern is somewhat surprising
given the importance of accurate pressure readings to well control. It is quite

possible that the crew had undergone a form of what Professor Diane Vaughan dubbed "the normalization of deviance," in which frequent deviations from standards prompt participants to redefine the standard to accommodate the deviations.[10]

THE DRILLING SIDE

Whereas a single vendor integrated the system controlling the vessel's position and ran it from a single location (with backup), the circulating system on the drilling side was an assemblage of components from multiple vendors. One set of components had been integrated into the *Horizon* when she was built, to which Sperry had added sensors and displays in ad hoc fashion when contracted to provide mud-logging services.

Crew members on the drilling side could not rely on a centralized display to control all the elements of the circulating system. When the marine side needed to change the level at which the *Horizon* floated, a crew member on the bridge would provide inputs to the Kongsberg computer console, which would initiate the necessary actions through pumps and valves. Sensors would read the new volumes in the relevant tanks, and the console would display those values.

When a driller needed to bleed pressure from the drill pipe, in contrast, he would dispatch a floor hand, who would go to one of the manifolds on the drill floor and turn a valve, and then report back that he had completed the task. The bridge ran like a modern vessel, where two or three individuals controlled everything through automation. The drilling side ran more like a nineteenth-century sailing ship, where those in charge exercised control by giving orders to others who then performed the necessary tasks.

Transocean's supervisor of the drilling side was called the *senior toolpusher*, who oversaw the thirty-eight Transocean employees whose duties required them to work with or around the drilling equipment. Transocean's administrative grades for the principal members of a drill crew were *toolpusher*, *driller*, and *assistant driller*. Each twelve-hour tour—6 a.m. to 6 p.m., or 6 p.m. to 6 a.m.—was staffed by one toolpusher, who acted as foreman of the crew, one driller, and two assistant drillers.

Adding to the complexity of the command problem, the rest of the drilling side was composed of ten individuals from eight other corporations, many of whom used equipment provided by their employers. The most significant

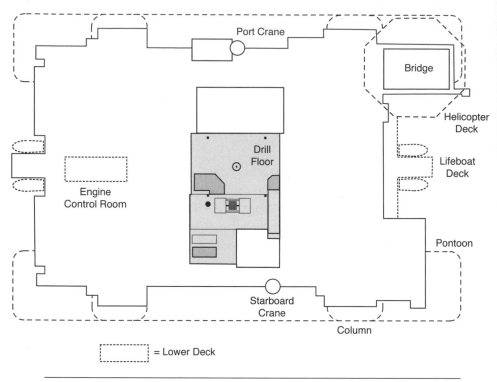

Figure 11.3. Layout of the main deck, showing the location of the drilling side.

of these was Halliburton, which did the cement work, and whose Sperry subsidiary furnished the *mud loggers* who monitored the circulating system in support of the drill crew. One mud logger was on duty at a time, and changed tours at 6 a.m. and 6 p.m.

The members of the drilling side were not required by law or federal regulation to hold licenses from a recognized maritime state or to fulfill any special educational requirements. Those who controlled or supervised the control of drilling equipment, as well as BP's well site leaders, were required by MMS to attend a one- or two-week well-control school every two years.

The drilling side occupied the drill floor, a raised area in the center of the vessel, as shown in Figures 11.3 and 11.4.

Elements 1 and 2 were the *main drill shack* and the *auxiliary shack* where an assistant driller was located. Shown outside them on the drill floor were the various *manifolds* whose valves controlled how the circulating system was lined up.

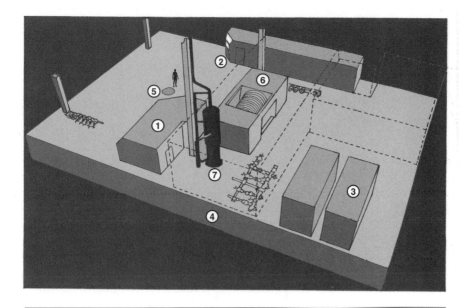

Figure 11.4. The drill floor.

Element 3 was the *mud logger's office,* a small structure separated from the drill shack by the structure housing the draw works. This office and the equipment in it used by the mud logger were provided by the Sperry subsidiary of Halliburton.

Element 4 was the machinery space under the drill floor. The piping for the circulating system ran through this space. The cement unit—where the senior WSLs were in the habit of going to observe a negative test—occupied the area under the mud logger's shack. The area under the drill shack was the shaker room, which held machinery to separate cuttings from mud before it went into the pits, which were one level below the shaker room.

Element 5 was the *rotary table,* which formed the open entry point to the well. Element 6 was the draw works, and Element 7 was the *mud-gas separator,* whose rupture and subsequent explosion probably killed the ten crew members who died on the drill floor or in the machinery space below.[11]

The interior of the drill shack, colloquially known as the "doghouse," is shown in Figure 11.5.

Element 1 was the *A-Chair,* usually occupied by the toolpusher. Keypads and joysticks on the arms of the chair controlled the drilling equipment. In

Figure 11.5. The interior of the drill shack. (1) A-Chair. (2) CCTV display. (3) BOP control panel.

front of the chair was the NOV HiTec console, which displayed the status of the circulating system on two screens. The freestanding display to the left of the A-Chair was the Sperry Sun display, which duplicated what the mud loggers saw in their office. It provided redundant and in some cases more accurate information than the HiTec system, particularly about flow. The windows in front of the chair faced the drill floor and the rotary table.

Next to the A-Chair was the *B-Chair,* where a driller sat. The person in the *C-Chair* diagonally across the drill floor was usually an assistant driller. It was Transocean's policy that two people should be in the drill shack at all times, and that one of the three chairs also be occupied at all times.

Between the two chairs was a display for the Kongsberg gas and fire detection system. The drillers seldom, if ever, interacted with this equipment.

Element 2 was a closed-circuit TV display that could be switched to cameras spotted around the rig, including one that was trained on a part of the circulating system down in the machinery space where the flow from the riser came out in the open. There was no camera covering the interior of the bridge.

Element 3 was the *BOP control panel,* containing a large number of switches, lights, and buttons that operated the BOP and displayed its status.

Crew members could perform emergency disconnect functions from this or a separate panel in the drill shack.[12] It was the practice on the *Horizon* for a subsea specialist rather than a driller to operate the BOP panel in normal conditions. This somewhat unusual arrangement may have led to delay or confusion in the last crucial minutes before the blowout, when no subsea specialist was in the drill shack.

INTERACTIONS BETWEEN THE SIDES

Supervising the master, chief engineer, and senior toolpusher was the offshore installation manager, or OIM, who reported to the Transocean rig manager in Houston. The single OIM generally worked the day tour and was on call at night. He held an international license issued by the Marshall Islands that required several months of specialized training.

According to Transocean policy, the OIM was in charge when the *Horizon* was latched up and acting like a drill rig, and the master was in command when the *Horizon* was moving and acting like a seagoing vessel—or in a state of emergency. The Republic of the Marshall Islands allowed this division of responsibility owing to a "clerical error" that had *Horizon* categorized as an anchored, stationary rig.[13]

The result was uncertainty about who was in charge after the blowout.[14] That uncertainty arose at least in part because written Transocean procedures implicitly defined an emergency as a "drift-off," in which a vessel cannot maintain its position over a well. There is no evidence that the procedures covered a situation in which the *Horizon* was latched up over a blown-out and burning well.[15]

Response to the blowout was further complicated by the fact that the bridge and the drill shack were not in direct visual contact—crew members communicated by telephone or handheld radio. Because there were no cameras in either the bridge or the driller's shack, the only way that one side could tell that the other was in difficulty or facing an incipient emergency was through a radio or telephone call.

This reliance on explicit notification contributed to confusion after the blowout by preventing the marine side from taking pre-emergency steps, such as issuing a "pan-pan" distress call,[16] or having crew members prepare for evacuation. Instead, the first indication the bridge crew had that Macondo had blown out was when the sudden buoyancy of the gas-filled riser jolted the rig.

This, then, was the Deepwater *Horizon* at thirty minutes to 1 a.m. on April 18, when crew on the drill floor held a pre-job safety meeting before running the first length of production casing into Macondo. The drilling side was a confederation of eight different corporations supported by an ad hoc, distributed assemblage of equipment from multiple vendors. Its design may have been state of the practice in the late 1990s, but by 2010 that design was obsolete and heavily dependent on the skill and attention of its crew to cope with a well as difficult as Macondo. And the *Horizon* was run down and in need of overhaul to rectify years of "run it until it breaks" maintenance.

Crew members on the drilling side had overcome the limitations of their equipment and organization on forty-nine previous wells because they were led by senior WSLs and performed the same sequence of drill, plug, and abandon almost every time. Now they were without their senior WSLs and facing the problem of executing an unfamiliar and, unbeknown to them, inherently dangerous set of procedures—a sequence of tasks that would expose the shortcomings of their combined human and technological system in the worst possible way.

CHAPTER TWELVE

Up to the Edge

OUR "LUXURIOUS RETROSPECTIVE POSITION" allows us to recognize transitions whose significance goes unnoticed at the time by those who experience them. The first half hour of April 18 was such a period. To us it represents the shift from planning to action, the beginning of an incremental loss of control that would end with the blowout, explosion, and fire on April 20. To the night tour on the drill floor and the aft deck, it was the start of a familiar rhythm under the bright lights of the rig. The difficult part would come the next day, when cementing would begin.

For now the casing lay on the aft deck, ready to be put on the conveyer and lifted to the drill floor, where it would be assembled stand by stand, finished off with a casing hanger, and lowered in place onto the wellhead. It was a Sunday, and the junior drilling engineer was on the rig, inquiring by email about regulations pertaining to negative tests, and later changing the displacement plan one more time. The senior drilling engineer was off-rig on a charity bicycle ride, and the Macondo project was essentially in the hands of oilfield service companies: Weatherford, which would assemble the sections of the production casing, having done some prefabrication ashore;

Dril-Quip, which furnished the running tool that held the casing as it was lowered into place; and Halliburton, which would do the cementing.

At this point the full function of the float collar enters the narrative. In Chapter 8 we described its partial function as that of a shelf that interrupted the passage of the wiper plugs as they proceeded down the production casing, permitting free cement to flow down into the shoe track, out the shoe, and up into the annulus between the production casing and the formation.

A second function of the float collar is to prevent external pressures from causing the liquid cement slurry to flow back up the casing. If this were all that was required, the float collar would be a simple and robust component of a larger system. However, designers added a third function—to selectively allow upward flow—and the result was a more elaborate—and in the case of Macondo, uncertain—device.

Figure 12.1 shows the logic of the float collar's operation.

Step 1 shows the float collar allowing upward flow of drilling mud into the production casing as that casing is run into the well. The situation here

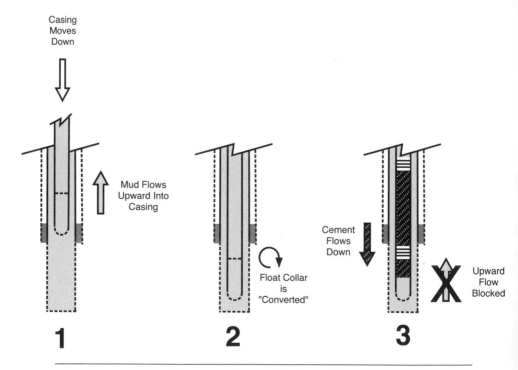

Figure 12.1. Logic of the float collar.

is like inserting a straw into a narrow-necked bottle. If the straw is open at the bottom, the contents of the bottle will flow into it in relatively uninterrupted fashion. If it is closed, it acts like a rod, forcing fluid up as it goes down. The temporarily open valve in the float collar permits the mud in the well to enter the casing smoothly, and without causing pressure surges that could break down the fragile formation of Macondo.

At Step 2 the float collar is *converted:* the driller uses the circulating system to generate a very specific combination of flow rate and pressure that, in systems terms, sends a command to the float collar sitting three and a half miles below the driller.

After the internal state of the float collar transforms in response to the command, the collar acts as shown in Step 3. Here, as described in Chapter 8, the collar has stopped the bottom wiper plug, which has ruptured, allowing the cement slurry to be pushed through both of them into the shoe track. At this point the collar will also prevent the liquid slurry from being pushed back up the inside of the production casing by mud entering it from below.[1]

As the crew was running the casing, the float collar was in the configuration shown for Step 1, open to mud coming in from below. It was the next step, conversion, which occurs between the setting of the casing and cementing, that raised the float collar to the status of a major risk factor in the production tail activity.

CENTRALIZER UNCERTAINTY

Our knowledge of the configuration of the production casing comes from two documents: a casing plan produced by BP and a casing tally produced by Weatherford.[2] Casing tallies are important documents in the transition of a well from temporary abandonment to full production, for they inform the crew doing that transition of the exact configuration down in the well. Errors in casing tallies can cause anything from stuck tools to an unusable well.

The BP plan and Weatherford tally show that four centralizers—rather than the six specified in the test and displacement plan—were installed on the production casing. The technician who assembled the casing, who said he thought "four or five" were on it, reinforced that tally, but the daily drilling report contradicted it.[3] All reports on the *Horizon* blowout we have examined show the possibly erroneous number of six centralizers in the well.

The confusion over how many the crew actually installed may reflect the inexperience of the junior drilling engineer, who may have specified an impractical arrangement with a centralizer next to both the shoe and the float collar. Someone, either the senior drilling engineer or a Weatherford employee, might have noticed this, eliminated those centralizers, and adjusted the locations of two others. Whatever the number of centralizers, the uncertainty associated with the configuration of such potentially important items further illustrates that discipline on the *Horizon* was slipping as the pressure to finish rose.

RUNNING THE PRODUCTION CASING

By 2:00 p.m. the Transocean crew had run all 5,600 feet of 7-inch casing (after taking an hour to deal with a bad casing joint) and switched their equipment to install the 7,400 feet of 9⅞-inch upper casing. That entailed installing the casing hanger that would hold it to the wellhead, and lowering the casing down the 5,000-foot riser on the landing string. On the way down, the string struck the ledge at 18,240 feet that had earlier blocked the wireline tool, as evidenced by a slight bobble in the record of weight on the draw works. By 1:30 p.m. on Monday the casing was in place, and the crew felt it was ready to begin cementing.[4] Nothing in their records or testimony indicates they thought the task of running the casing had been anything other than uneventful.[5] The uncertain physical state of the casing material did not arise in any depositions or trial testimony.

Late that night or early the next morning, Schlumberger employees arrived on the rig, prepared to perform the cement bond log should any questions arise regarding the success of the cement job.

CEMENTING

The signal from the circulating system that the float collar used at Macondo would recognize was 500 to 700 psi of downward pressure at six barrels per minute of mud flow. A distinctive and well-known pressure and flow profile from the circulating system would then signal a resumption of normal mud flow and thus a successful conversion of the float collar—values MI-Swaco provided from a computer simulation at its Houston office.

A Weird Situation

With the junior drilling engineer and the substitute WSL present, the drill crew ramped up the pumps to convert the float collar. However, the signals from the circulating system did not show any flow at all through the device. The crew bled off the pressure and tried seven more times, each time increasing the pressure applied to the float collar, and each time bleeding that pressure back to zero when no flow was observed.

The junior drilling engineer and the substitute WSL made calls to BP Town and vendors seeking guidance, and finally on the ninth attempt at 3,142 psi, got an indication of flow through the device. However, it and the associated pressures were lower than predicted by the MI-Swaco model, and flow ramped up more quickly than normal for that class of flow collar.[6]

Models such as those produced by MI-Swaco are based on assumptions, and in this case the model may well have assumed that the float collar was being converted in a "normal" hole, not the ragged, washed-out one of Macondo. The failure of the flow to conform to that predicted by the MI-Swaco model could be interpreted as a signal that the cement job was taking place in unusual circumstances—a signal that did not appear to have changed the crew's determination to press on.

The junior drilling engineer heard the crew conclude that the standpipe pressure gauge was wrong, and that "this is standard out there."[7]

Yet the cementer on duty at the time testified that "some concern was shown by many people on the drill floor" and he himself was concerned enough to call the Halliburton representative in Houston to report the situation, who told the cementer to "keep him informed of any decisions that were being made from then on in."

The cementer also heard the substitute WSL say: "I'm afraid that we've blown something higher up in the casing joint."[8]

The junior drilling engineer later emailed in response to a question: "Yah we blew it at 3140, still not sure what we blew yet."[9]

At 6:20 p.m. the cementer wrote in his tally book: "Shut down while BP decides what to do next."[10]

A BP interviewer's notes on a post-blowout interview with the substitute WSL indicate that he told the incoming junior WSL, "This is a weird situation."[11]

Pressing On

Despite all these reservations, after thirty minutes of consultation by various parties on the rig and in Town, the well team leader issued the order to continue the cement job. The crew circulated into the well 110 barrels of clean mud[12]—an amount calculated to clean the bottom where the cement would go without putting the kind of pressure on the formation that would occur during a full bottom-up circulation. Several experts criticized this decision after the fact because the system would not have circulated enough mud to prevent contamination of the cement by debris and mud that had gelled during sustained exposure to high temperatures at the bottom.[13]

The rest of the preparation and pumping of the foamed cement occurred without incident. The pressures and flow signals from the circulating system corresponded to those calculated in advance. In accordance with the decision chart produced during the MOC meeting, BP sent the Schlumberger crew that was prepared to perform the cement bond log back to shore on the next helicopter.

At 2:52 a.m. on Monday, April 20, the junior drilling engineer emailed BP Town that the cement job had gone well, and he went to bed.[14] The next night neither he nor anyone associated with the *Horizon* would get much sleep.

CHAPTER THIRTEEN

The Final Day

THE LAST DAY in the life of the *Horizon* and eleven of her crew featured un-coordinated actions driven by disaggregated information. It began with the confusion within BP and between BP and the Transocean crew about the sequence of tasks to be performed, as described in Chapter 9, and continued in the same manner to the very end.

After Transocean employees somewhat grudgingly accepted the sequence of tasks agreed on earlier by the BP well team leader and the junior drilling engineer, the crew settled down to complete four tasks before cementing the second, upper plug in the well. The first was to perform a positive pressure test on the production casing—the only well integrity test then mandated by regulations.

The test entails using the cement pumps to increase pressure in the pro-duction casing by 2,500 psi, and observing whether the pressure holds for thirty minutes. The test serves only to ensure that fluid cannot leak out of the casing. It puts no stress on the cement at the bottom of the well, so it cannot determine whether hydrocarbons can leak in. The crew uses a negative test to determine that.

Data from the Sperry Sun system suggest that the production casing passed the positive pressure test.[1] However, about four hours later, when the crew began the first displacements of spacer and seawater, some fluid began flowing out of the production casing and into the formation—contradicting the results of the positive pressure test. These lost returns continued through the second displacement until the well became unbalanced and hydrocarbons began flowing into the production casing.[2]

This leaves us with one of the central puzzles of the blowout, one that any acceptable explanation of what happened at the bottom of the well must solve: What changed in the well between the positive pressure test and the first displacement? We will probably never know.

One thing we do know: on the final day the crew did not proceed with the methodical wariness that the senior WSLs practiced—a wariness that undoubtedly had protected the *Horizon* on previous wells. Those senior WSLs would perform well-control calculations before dangerous tasks such as negative tests so that they could recognize unusual signals from the circulating system during those tasks. They treated the negative test with special care. As one of them testified: "Before every negative test, we put down step by step what we're going to be doing, how many strokes we're going to be pumping here and there. And we'd give that to the driller and the toolpushers and let them look at it and make sure everybody agreed what we were going to be doing once he started."[3]

This exercise, in effect, fleshed out the one-line "perform negative test" instruction in the procedures that the WSLs typically received from BP Town. Further care was taken: "And once the mud engineer and myself, the driller and the toolpusher got finished with [the step-by-step procedure], then we did a task-specific THINK drill on it to involve the rest of the crew so everybody would be familiar with what we were doing."[4]

But these crucial steps were not embodied in formal procedures—they were lore, resident primarily in the minds of the senior WSLs. There was no mechanism to document, teach, or enforce them when the senior WSLs were absent from the rig. Motivated by some unknown combination of haste, distraction, and reassurance provided by the successful positive test, the WSLs and the senior Transocean crew on board that fatal day omitted those steps, leaving the crew with nothing but a footnote on the mud engineer's one-page procedure for the displacements: "NOTE: Good communication will be necessary to accomplish a successful displacement. If you are not sure, stop and ask."[5]

Nobody did.

THE LAST FEW TASKS

At 11:30 a.m. the junior drilling engineer sent an email to the substitute WSL[6] with the final version of his displacement procedure and then boarded the helicopter that would take him to shore. This procedure, like the others, consisted of one-line descriptions of a sequence of steps. The sequence defined a partial displacement from 8,367 feet to above the wellhead, and a negative test using the kill line that, as we described in Chapter 3, bypassed the annular preventer and was given a role in some, but not all, of the procedures for a negative test that had circulated in BP Town and between Town and the rig. The crew was not accustomed to using the kill line for negative tests; they usually tested using the drill pipe instead. It is not clear how this procedure may have influenced later events, but it is another indication of the lack of a uniformly understood set of steps for conducting this critical test.

Offloading Mud

Around 1:30 p.m. the *Bankston* used her dynamic-positioning abilities to assume a station on the port side of the *Horizon*. Crews ran a hose between the two vessels and started pumps to transfer used mud from the *Horizon*'s pits over to the *Bankston,* a necessary step before moving the rig to the Nile well and then on to Kaskida.

The configuration of pits and auxiliary pumps on the *Horizon* meant that the transfer required complicated movements of mud from pit to pit before it was offloaded. During this time the crew also began cleaning pits and emptying settling tanks full of sand separated from returning mud. All this activity made it impossible for those on board, as well as later forensic analysts, to directly determine the changes in pit volumes, which are fundamental to monitoring well control. In fact, the day tour mud logger—who, like her night tour colleague, was charged with being a second observer of well-control parameters—was so concerned about being deprived of volume measures that she arranged for an assistant driller to call and advise her of the steps the drilling crew was performing, presumably so that she could use other, inferior signals such as flows and pressures to perform her well-control responsibilities.[7]

Visitors Arrive

At about 2:30 p.m. the helicopter carrying high-ranking BP and Transocean visitors arrived at the rig. The BP side included the vice president of drilling and completions—the most senior executive with direct oversight of BP Town, who had served four months in that assignment. With him was the operations manager, newly promoted over the well team leader. The Transocean side included two operations managers, one of whom supervised the *Horizon*'s rig manager, and the other who was a replacement, added to balance the party.

The OIM later testified that the *Horizon* was used to visitors, and that their presence was not a distraction.[8] However, evidence suggests that on at least one occasion the presence of VIPs inhibited crew members from admitting they were having difficulties with the well. At the very least, both the OIM and the senior toolpusher were occupied with the visitors and unavailable to compensate for the absence of the senior WSLs.

Preparing for the Negative Tests

Around 3 p.m. the mud engineer briefed the crew on the drill floor on his final displacement procedure. The crew had already run drill pipe down to 8,367 feet, the level at which the upper cement plug was to be placed, and lined up the four main pumps to conduct the negative test and the displacements. Two pumps were lined up on the drill pipe and one on an auxiliary line to the bottom of the riser called the boost line, which was used to accelerate the return flow up the riser. The fourth pump was attached to the kill line.

After the briefing, the crew began pumping seawater into the auxiliary lines of the BOP to remove mud in them. The boost line was cleared first and without incident. Immediately after running the pump to fill that line with seawater, the crew ran the kill line pump for less than a minute, causing pressure on that line to spike to 6,000 psi (Figure 13.1).

An independent analysis of well hydrostatics commissioned by Transocean attributed this event to a test of surface lines on the drill floor: "Note

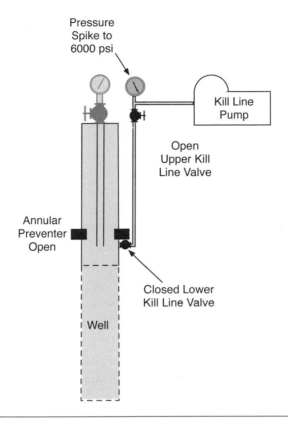

Figure 13.1. The pressure spike.

that the spike in kill line pressure between 15:17 and 15:19 was a planned pressure test of the choke manifold lineup prior to displacing the choke and kill lines. This event will not be analyzed in further detail in the discussion of surrounding events."[9]

We can find no independent evidence to support this conclusion. Sperry data suggest that the crew had used the cement pumps an hour earlier to raise pressures in the choke, kill, and boost lines to 3,000 psi. The crew would seemingly have had no reason to use a main pump—an inappropriate device—to produce much higher pressure.[10] During the last hour, a similar spike put the kill line pump out of commission, making it even less likely that the crew had deliberately used a main pump to produce higher pressure.

Other drilling experts interpret the Sperry data as suggesting that the crew had inadvertently started the kill line pump while the lower kill line valve was closed, as shown in Figure 13.1.[11] If so, then the crew was inattentive to that valve or the BOP control, for it was subject to intermittent failure. Whatever the cause, repeating the mistaken impression that the valve was open when it was closed could partially explain the crew's fatal misinterpretation of the second and final negative test.

The crew then pumped in 450 barrels of the viscous, LCM-based spacer to separate mud from seawater they would pump in next. This was, as we explained in Chapter 9, the act that BP investigators later called "sham recycling,"[12] the use of a material in a way no one had used it before solely so that it could be legally discharged into the ocean.

The crew then pumped about 350 barrels of seawater into the well to raise the spacer above the BOP. It was sometime during these two displacements that it is likely that something happened down in the well to invalidate the results of the positive pressure test and open a path from the inside of the production casing to the formation. That casing was two and one half miles of tubular steel, hanging vertically, fastened to the wellhead at the top and cemented at the bottom, and filled with oil-based drilling mud. The displacements replaced hot drilling mud—210 degrees Fahrenheit at the bottom—with cooler spacer and seawater, raising the possibility of contraction of the production casing and subsequent stress on cement and fittings.[13]

Whatever the cause, we do have direct and indirect evidence, from the Sperry data for this period and later, that mud began flowing out of the production casing and into the formation at this time. However, neither the crew members in the HiTec chairs nor the mud logger could have discovered these lost returns by observing changes in pit volumes, because, as noted, other crew members were moving mud from pit to pit before transferring it to the *Bankston*.

None of the procedures in the trial evidence mentions this transfer, and no one testified as to why the crew started it when it did. But that transfer obscured two pivotal facts: that the production casing was open to the formation, and that they were in the middle of a well-control situation. It was time to stop. Instead they proceeded, unaware.

In addition, it is probable that the mud was being lost from the bottom of the production casing while seawater was being pumped in the middle, as shown in Figure 13.2. This meant that a large amount of the heavy, viscous spacer would not have been where it was supposed to be. Instead of being up above the annular preventer, so that the negative test would be based

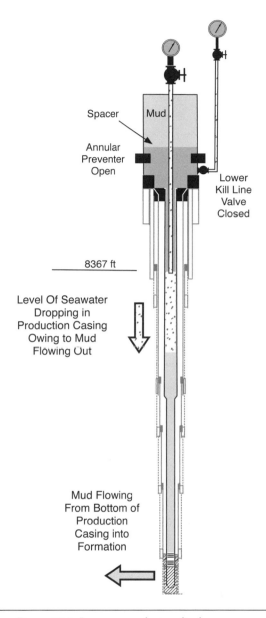

Spacer

Mud

Annular
Preventer
Open

Lower
Kill Line
Valve
Closed

8367 ft

Level Of Seawater
Dropping in
Production Casing
Owing to Mud
Flowing Out

Mud Flowing
From Bottom of
Production
Casing into
Formation

Figure 13.2. Lost returns during displacement.

on a homogeneous column of seawater, as much as 3,000 feet of it was below that point, confounding the negative tests in ways no one could predict at the time or analyze convincingly afterward.

THE NEGATIVE TESTS

The crew prepared for the first negative test by closing the annular preventer and lining up the circulating system as described in Chapter 6. This arrangement moved the center of control from the drill floor to the cement unit, whose pumps, pressure gauges, and test tank would be used to detect flow. In the first of many deviations from the "way [the senior WSLs] did it," the substitute WSL did not go down to the cement unit to monitor the test, relying instead on information relayed to him from the cementer.

At this point the crew continued the improvisations that began when the well team leader and the junior drilling engineer decided to displace before testing rather than test first, and that continued when BP overrode the objections of its own senior drilling engineer and members of the Transocean crew at the last minute. The effect was to replace an agreed-upon although undocumented series of actions on previous wells with something of the same name—something that not every participant understood in the same way.

As we noted in Chapter 6, survivors and outside experts have disagreed on how many negative tests the crew actually performed, whether one step or another actually constituted a negative test, and what constituted success or failure for some steps.[14] In the interests of clarity we shall simply describe each action the crew took, with the understanding that the exact sequence and participants in any given action are uncertain.

Despite this uncertainty, the crew's actions were consistent in one regard: they represented a persistent search for a green light indicating a successful cement job—a search that continued despite flickers of red along the way. The only criteria for a green light the crew accepted from the once-integrated "way [the senior WSLs] did it" were thirty minutes of no flow from the well and no pressure on whatever line crew members were watching, whether the drill pipe or the kill line. Those two absences would indicate that the cement at the bottom of the well was keeping hydrocarbons out of the production casing. This turned out to be a tragically narrow view of what constituted a green light, for an absence of flow and pressure could have other causes.

The First Negative Test

This test was configured in the traditional fashion for the *Horizon,* with the drill pipe monitored in the cement unit. It began with a missing step: there is no evidence that anyone calculated how much mud should have come out of the riser when the cementer opened the valve in the cement unit to bleed off residual pressure generated by pumping in the 350 barrels of seawater. A Transocean calculation after the blowout showed an expected value of slightly more than three barrels.[15]

When the cementer on station opened the valve, he observed a flow of about twenty-three barrels before he got the order to shut the valve. The cementer, who had assisted the senior WSLs with several negative tests done their way, concluded at the time—and maintained consistently afterward—that this flow indicated a test that had failed in its very first step. He waited at his station for word to that effect to come from the drill floor.[16] None came. Instead, without involving him, the substitute WSL and the Transocean drillers began to discuss what could have caused that flow—a clear indication that they did not view it as an indication of a failed cement job.

It is not clear whether the junior WSL was on the drill floor during this activity. He was, however, in his office immediately afterward, where he talked with the visiting BP operations manager, who had broken away from the tour to discuss the anomalous flow of mud with him. No evidence sheds light on the substance of this conversation. The operations manager then rejoined the tour.

At 5:17 p.m. the mud engineer who had devised the detailed displacement procedure ordered—evidently on his own initiative—the crew to stop the transfer of mud to the *Bankston.* He later testified that he gave this order after he heard about the anomalous flow at the first attempt to bleed off pressure at the cement unit; "I wanted to be ready in case whatever direction we went."[17]

In an interview with BP investigators, the junior WSL asserted that he had ordered the crew to stop the transfer,[18] but no other testimony mentions this order. No one informed the day tour mud logger that the transfer had stopped.

Confusion Reigns

It was now approaching the time for crew changes for individuals working the 6:00 p.m. to 6:00 a.m. tour, and crew members began drifting into the drill shack to conduct the usual handover conversations with the people they were replacing, making the shack a fairly crowded place. By the time the dignitaries arrived around 5:30 p.m., the shack held the day and night tool-pushers, the day and night subsea supervisors, two mud engineers, and three or four drillers and assistant drillers. The junior WSL also showed up, as did a trainee WSL, whose duties consisted primarily of observing how things were done.

Over at the mud logger's office, which was located some distance from the drill shack, the day tour mud logger was also handing off oversight to the night one. Because no one had told her that the transfer of mud to the *Bankston* had stopped, she informed the night mud logger that it was continuing and he did not discover the truth until after the blowout—another example of the growing disarray during the displacement activity.

The four arriving VIP visitors were accompanied by the OIM, the senior toolpusher, and the chief engineer. When they got to the drill floor, it was surrounded by red tape to signify that a critical operation was under way and entry was by permission only. The senior toolpusher obtained permission and the group entered the drill floor, took a quick look into the crowded drill shack, and retired to the area behind it, where they discussed personal safety procedures. One of the Transocean managers on the tour lingered at the drill shack long enough to hear the crew discussing the abnormal flow. As the day subsea supervisor stated later, "We had a pretty good task going on."[19]

The visiting Transocean manager asked the OIM and the senior tool-pusher to stay behind and help with what the senior toolpusher later characterized as "a little bit of a problem."[20] The chief engineer assumed the duties of tour leader, taking the VIP group off the drill floor and down into the pontoons.

Just after they left, someone noticed that the fluid level in the riser—which crew members can observe by looking down into it from the drill floor—had dropped below its normal level. The crew topped it off with a volume of drilling mud estimated later between twenty-five and sixty barrels.[21] The substitute WSL was making his way to the drill floor at this time and observed afterward, "I don't know why they were filling the riser—they were topping it off. One guy was watching it with a flashlight."[22]

The crew attributed the loss of fluid to a leaking lower annular preventer. The day subsea supervisor increased the pressure with which it squeezed the drill pipe and later observed, "With 3 kicks on the well, the pressure in the lower annular was iffy at best."[23] A more likely circumstance was that the fluid loss occurred before the annular preventer was closed, and was the result of mud leaving the production casing and going into the formation. It is conceivable that, owing to the success of the positive pressure test, no one in the crew considered that possibility.

Whatever the cause, the drop in level was a sure indicator that the viscous spacer was now back down in the upper portions of the well, where it could confound pressure and flow readings. One of the senior WSLs noted,

> But if I was onboard and we lost mud back below the annular, first thing I'd do is open the annular and replace all the water back down to the drill pipe and move the spacer back to where it's supposed to be.
>
> And then I'd close both annulars, make sure we didn't have no leak and try to do the negative test again. But once this annular leak got that heavy fluid back below the annular, your whole system's screwed up. You can't tell what's going on.[24]

No one closed the annulars or checked whether more fluid was being lost. Instead, in a textbook example of go fever, the crew adopted a variety of reassuring explanations for the loss of fluid and pressed on.[25]

The drill crew then made another attempt to bleed the drill pipe down to the desired zero pressure and no flow, and again observed pressure and flow. Around 6:00 p.m. the substitute WSL, the senior toolpusher, and other Transocean employees discussed whether to try to achieve thirty minutes of no flow and no pressure on the kill line instead of the drill pipe. Some Transocean people evidently resisted yet another deviation from their traditional idea of a negative test. The substitute WSL prevailed, at least in part by noting that the procedure approved by MMS had called for conducting the test on the kill line.

This exchange again reinforces the level of confusion regarding the tests. There was no physical difference between a test on the kill line and a test on the drill pipe. The two lines were like two straws inserted in the same soda bottle: just as each straw would experience the same degree of carbonation, each line would experience the same pressure and flow.

The argument that the tests should be brought in line with the MMS-approved procedure was also specious. BP had already violated that procedure by moving the tests from before the first displacement of mud with seawater to after, and it was too late to change that.

The Second Negative Test

After this discussion the senior toolpusher went down to the galley to eat
dinner. The drill crew then ran one of the pumps for a few strokes to ensure
that the kill line was full of seawater, and connected the kill line to a small
tank on the drill floor not monitored by the Sperry Sun equipment. While
all this was happening, the pressure in the drill pipe, as monitored in the
cement unit, slowly rose to 1,400 psi and stayed there. The drill crew tried
several times to bleed off pressure from the drill pipe, and each time it came
back up.[26]

A long and controversial discussion ensued in the drill shack about the
1,400 psi reading on the drill pipe. Various accounts note that the participants included both WSLs, the day and night toolpushers, the night driller
and two assistant drillers, two mud engineers, and the trainee WSL. The
latter testified that the senior toolpusher was there, but he insists he was in
the galley eating dinner. Of the remaining group, all perished save the day
toolpusher, the two WSLs, and the trainee WSL. The day toolpusher gave
one account to Transocean investigators in which he did not mention being
in the meeting, and the two WSLs gave only abbreviated statements to BP
investigators before asserting their Fifth Amendment rights.

We are then left with only the various testimonies of the trainee WSL
from BP. This individual stated that the Transocean night toolpusher ascribed
the 1,400 psi to something called the "bladder effect" or "annular compres-
sion," in which the weight of the mud in the riser transmits a downward
force through the closed annular barrier, which in turn generates a U-tube
pressure back up through the drill pipe.

One important question is why this critical discussion and its erroneous
and disastrous conclusion occurred within the confines of the drill shack,
with no recourse on the part of either Transocean or BP employees to outside assistance. This is particularly surprising in the case of BP, as one drilling
expert viewed its method of rotating WSLs (half a hitch on day, half a hitch
on nights) as a policy of "drilling from shore," or limiting the initiative
exercised by the WSLs.[27] Without further testimony we will probably never
know why the crew did not ask for outside help.

A subsea supervisor who had shared the drill shack with the Transocean
night toolpusher during many tours suggested after the blowout that the tool-

pusher was inhibited from asking for help by the fact that the people he would have called upon were in a room with the VIP visitors from BP and Transocean: "On a normal operation, [the night toolpusher] would have contacted the OIM [or] senior toolpusher; but the conference—the phone in the conference room was, I mean, as soon as you call, it was like a speakerphone, and I felt he had too much pride to call and said he was having issues."[28]

In another interview, the subsea supervisor added that "[he] was the type that carried the whole rig on his shoulders."[29]

At about 7:45 p.m. the crew checked the tank on the drill floor, observed that it had taken no flow, and pronounced the negative test a success. Forensic analysts have struggled to come up with a purely hydrostatic explanation for a situation where both the drill pipe and the kill line were connected to the same pressure vessel but one reported 1,400 psi of pressure and the other zero. The simplest and most likely explanation is that the kill line was blocked by either viscous spacer or an inadvertently closed—and unreported by the BOP control system—lower kill line valve. This possibility evidently never occurred to the crew. They had finally seen the green light they had been looking for, and developed a benign explanation for the 1,400-psi red light shining at the same time.

In the Grip of Go Fever

The substitute WSL went to bed, and the cementer who had witnessed the earlier flows from the drill pipe came up to the drill floor to see what he should be doing next. He testified that he "was instructed by the driller and toolpusher that they had achieved a successful negative test on the rig floor and to go ahead and get our job procedure ready for the surface plug."[30]

The drill shack then filled with people for a pre-job meeting before the second displacement.[31] At two minutes past 8:00 p.m. the night subsea supervisor opened the annular barriers in the BOP, and the driller in the A-Chair started the pumps to displace the remaining mud out of the riser. Now only the column of mud in the well and the riser separated the *Horizon* from The Edge.[32]

On the evening of the blowout, in one of several simultaneous activities on the rig, a work party on the aft deck was preparing equipment to set the lockdown sleeve. One of its members, a senior casing specialist, was asked later if he was worried about the outcome of that job, and in response he reflected on the moods he had encountered on a rig at that stage of a well:

> You know, we've seen so many casing jobs over the years, when you get to that point, everybody goes to the mind set that we're through, this job is done . . . when you run that last string of casing, and you've got it cemented, it's landed out, and a test was done on it, then you say, "This job, we're at the end of it, everything's going to be okay."
>
> Now I'm telling you this not from a supervisor, not from the well site leader's office, but from the working men, that are out there—we finished this well, you're thinking ahead to your next job, you're moving on.[33]

And it is likely that the thought of moving on to Nile and Kaskida, already the subject of the Sunday safety drills, was not far from any of the crew's minds.

CHAPTER FOURTEEN

Going Over

AT 8:06 P.M. the first of the two drill pipe pumps pushing seawater into the well and mud up the riser for the second displacement was coming up to speed. Ten minutes later the pump was there, and the crew started the second drill pipe pump. Eight minutes after that the crew started the boost pump to speed the flow of mud up and into the rig's storage pits.

The pumps were moving the heavy mud—whose weight formed the only barrier between the *Horizon* and the pressurized and explosive hydrocarbons of Macondo—out of the riser at a theoretical rate in excess of 1,250 gallons every minute, and replacing it with lighter seawater. In forty-five minutes, again in theory, the weight of the fluids in the well and the riser—seawater, spacer, and mud—would be insufficient to hold back those hydrocarbons, and Macondo would kick for the last time.

As with so much else about Macondo, the theoretic model did not fully match reality. Just as during the first displacement, mud was flowing out of the bottom of the well and into the formation while the pumps moved seawater into the middle of it. The Sperry data show a growing difference between the amount of fluid pumped into the riser—calculated from the number of pump strokes—and the volume of fluid in the storage pits. There is no evidence that the crew on the drill floor noticed this discrepancy. In an interview with a BP investigator, the mud logger on duty at the time said he was not monitoring pit volumes because he thought mud was still being transferred to the *Bankston,* evidently because no one had told him it had stopped.[1]

THE FINAL HOUR

From this point until the blowout, events on the *Horizon* unfolded in a confused and uncoordinated fashion, accompanied by extreme disaggregation of information—events that in some cases defy analysis, and as a set possess no structure save the order in which they happened.

8:45 P.M.
Down in a lower-deck conference room, the VIP visitors from BP and Transocean had wrapped up their meeting with the OIM and senior toolpusher and were headed to the bridge to see how it operated. The last topic of the meeting was a question from the BP vice president: "Why do you think this rig performs as well as it does?"[2]

8:52 P.M.
The crew, for some unknown reason, cut the rate at which the pumps were operating almost in half. This lowered the hydrostatic pressure in the well, causing the outward flow of mud from the bottom to level off and fluctuate for a few minutes before turning around and becoming the influx of hydrocarbons that would become a blowout. As before, there is no evidence that the crew in the drill shack noticed the change in the difference between the amount of seawater pumped in and the amount of mud flowing out. The mud logger was still under the impression that he did not need to track the volume of mud in the pits because it was being transferred to the *Bankston.*

By grim coincidence, as loss started to become influx, the senior drilling engineer in Town called the junior WSL on the rig and spoke for ten minutes.[3] The nature of the call is uncertain because both participants exercised their Fifth Amendment rights during the trial. However, notes from BP's internal investigation show that the anomaly during the negative test of 1,400 psi of pressure in the drill pipe and zero in the kill line entered the conversation. The notes also show that the senior drilling engineer did not understand this anomaly enough—or view it as significant enough—to order the junior WSL to stop the displacement, shut in the well, and begin remedial action.[4]

The junior WSL then left the drill floor, telling the crew to call him in his office when it was time for the static sheen test.[5] The movement of the duty WSLs on and off the drill floor during the two displacements—leaving the Transocean crew alone for periods of time—was another deviation from the practice of the two senior WSLs:

Q. Explain to me again why you believe that it is important to be there for
    that entire period of time on the rig floor.
A. To complete displacement, you need to be there.
Q. And—and why is that, sir?
A. 'Cause the well can give up at any time, once you're reducing hydro-
    static. Even though it has passed negative test, it doesn't guarantee the
    well's going to stay intact.[6]

The junior WSL later told interviewers that he was on the drill floor up through the sheen test described below and left immediately afterward.[7] No evidence is available to resolve this discrepancy, but in any case no WSL was on the floor during the complex and ultimately deadly events that followed that test.

9:08 P.M.

An MI Swaco technician called a *compliance specialist* made his way to the mud engineer's office next to the shaker room to perform the sheen test called out in the mud engineer's displacement procedure. The compliance specialist had arrived on the *Horizon* just that morning, and in less than an hour he would be jumping off the lower deck into the burning ocean.

His duties were that of an internal inspector, tasked with certifying that anything discharged off the rig conformed to environmental regulations. He was to go to a bin in the shaker room where the mud coming out of the riser flowed in the open, and wait until he saw the distinctively colored spacer arrive, at which time he was to catch a sample and verify that it was free of oil.[8]

As with the negative test, no published procedure for a static sheen test appears to have existed. However, testimony indicates that the drillers were supposed to stop the pumps until the fluid passed the test, and only then divert the return flow of mud overboard. If the crew had coordinated the test with other activities, it could also have served as a flow test on the well and alerted them to the fact that they were in the middle a kick.

However, the Sperry data show that the drill crew not only stopped all three pumps but also lined up the valves in the shaker room to divert fluid coming out of the riser overboard, which meant they no longer had changes in pit volume available as a kick indicator. They also did not notify the mud logger of the new lineup. At this point the shortcomings in the circulating system we describe in Technical Note 6 in the Appendix caused it to appear to the mud logger as if there was no flow and appear later to the compliance specialist as if the flow was the result of the pumps running, while all the time the most likely circumstance was that the well was flowing because hydrocarbons were entering it.

9:10 P.M.

The drill crew started transferring mud between pits, as if preparing to resume transfer of mud to the *Bankston*—an indication that they likely thought things were proceeding normally.

9:13 P.M.

A phone call by the off-duty senior toolpusher reinforced the sense that nothing was wrong:

> I called the rig floor and I talked to [the night toolpusher]. And I said "Well, how did your negative test go?"
>
> And he said "It went good." He said "We bled it off. We watched it for 30 minutes and we had no flow."
>
> And I said "What about your displacement? How's it going?"
>
> He said "It's going fine." He said "It won't be much longer and we ought to have our spacer back."
>
> I said "Okay." I said "Do you need any help from me?"
>
> And he told me "No, man." Just like he told me before he said "I've got this." He said "Go to bed. I've got it."
>
> He was that confident that everything was fine.[9]

The statement that the spacer should have been arriving at the surface "soon" is consistent with the testimony of the compliance specialist.

9:14 P.M.
The crew restarted the two drill pipe pumps. Why they did this before the probable time of the sheen test is unknown.

9:16 P.M.
The crew restarted the boost line pump. Why the crew started the pumps while the equipment in the shaker room was lined up to send returns from the riser overboard is unexplained. The mud logger called the drill floor and asked about the staggered start of the pumps and was abruptly told "that's just the way we was doing it this time."[10]

9:18 P.M.
The compliance specialist likely took his sample of spacer for the static sheen test between 9:16 and 9:18 p.m. Possibly as part of that test, the crew stopped the drill pipe pumps while leaving the boost line pump running—a highly unusual act, and inconsistent with the fact that they had lined up the returns to go overboard.

The crew then stopped the transfer of mud between pits that they had started eight minutes earlier in likely anticipation of resuming the offloading of mud. The reason they stopped the transfer is unknown.

Then the crew attempted to start the kill line pump, only to have it produce a second 6,000-psi spike in pressure and then blow the pump's safety valve.

An assistant driller ordered the two crew members in the shaker room down to the pump room on the next lower deck to replace the safety valve and paged another assistant driller and floor hand to join them, where thirty minutes later they all perished. No credible explanation has been put forth as to why the drill crew started the pump or why the drill crew insisted on repairing it immediately.

The crew restarted the other three pumps. The boost line pump sped up—a possible sign that gas had entered the riser and reduced the load on it.[11]

9:30 P.M.
The crew stopped all three pumps. At this moment they had pumped within fifty barrels of the amount of spacer and seawater theoretically required to complete the displacement—and almost certainly knew that. It is very likely, then, that they stopped the pumps not as an emergency act but as part of a

plan. Because diverting the returns overboard deprived the crew of pit volume measurements, they were also deprived of a clear signal that possibly 500 barrels of hydrocarbons were in the well and moving toward them.[12]

At this moment three sets of signals could have alerted them to the peril they were in. The first of these was information from the Sperry system showing roughly 800 psi of pressure in the kill line and 1,200 psi in the drill pipe. As during previous tests, both lines were connected to the same body of fluid and should have formed a U-tube of equal pressure. A second set of signals would have come from the HiTec paddle-style flow sensor, whose issues we describe in Technical Note 6 in the Appendix.

A third signal would have come from someone in the shaker room, who would have directly observed flow and used the intercom to report that to the drill crew. If the crew in the drill shack had then used the intercom to summon one of the two crew members who normally would have been in the shack, they would likely have received no answer, as those individuals were involved in the hurried effort to get the pump on the kill line running again. One of the two floor hands, whose final location is unknown, may have been dispatched to the shaker room to observe if there was flow and perished there.

We know that at this time the crew in the drill shack was concerned about the signals showing different pressures in the kill line and drill pipe. The first mate of the *Horizon*—who was responsible for the cargo on the vessel—had gone to the drill shack to ask when the dry cement needed to set the final plug had to be moved from storage into the cementing unit.

He found the driller in the A-Chair and the night toolpusher standing next to him. They were occupied with the HiTec display and the first mate waited for them to finish. He described their demeanor as normal. The night toolpusher stated, "We are seeing a differential pressure" and added, "We may need to circulate."

The night toolpusher then informed the first mate that there would be a one- or two-hour delay in cementing the final plug.[13]

For some reason, possibly in response to the pressure anomaly, the driller dispatched one of the floor hands (who survived) to the main manifold with orders to open the valve on the drill pipe and bleed off its pressure. He did this, which equalized pressures in the kill line and the drill pipe at 800 psi. He then closed the valve, and the pressure in the drill pipe rose back to 1,200 psi. The Sperry data then show a quick jump in volume for the trip tank at 9:42 p.m., and then no further change—a sign that the crew had opened the valve to the trip tank for a flow test.

It was too late. The drill pipe pressure was dropping in the classic indicator of a kick. Gas was in the riser, and, as they say in the industry, Macondo was coming in at them.

9:45 P.M.

Sperry data show that the men in the drill shack had now closed the annular preventer and other barriers of the BOP in a desperate struggle with the well. In a move that was heavily criticized after the fact, the crew had selected the relatively fragile mud-gas separator (MGS) as the exit route for gases coming up the riser, instead of the more robust and direct overboard lines. The MGS was an environmental protection device that vented gas into the atmosphere but routed heavier fluids like mud back into the pits. The overboard lines vented everything over the side and into the Gulf, an act that was illegal except in an emergency. Why the crew chose the MGS is unknown, but they may have believed they were in the early rather than the last stages of a kick.

In these last minutes, life on the rest of the rig exhibited a fragile continuity. The *Bankston* hovered next to the *Horizon* on her port side, hose still connected, waiting for the order to offload the rest of the drilling mud in the *Horizon*'s pits.

The four high-ranking visitors were on the bridge, amusing themselves by playing with the dynamic positioning simulator and making small talk with the third mate. She was a young woman, just two years out of the California Maritime Academy, and was focusing on her duty station, monitoring the dynamic positioning equipment and the panel showing the status of the hundreds of gas and fire alarms distributed around the rig.

The OIM was taking a shower, and the senior toolpusher was in his cabin, having just phoned his wife. The substitute WSL was in bed, and the junior WSL was in his office doing paperwork. The first mate was making his way back to the bridge, and had stopped at the night subsea supervisor's office to chat with him and an off-duty assistant driller who was also visiting there.

BLOWOUT

Like many of its kind, the Macondo blowout arrived at the *Horizon* "on little cat's feet," with a small discharge of water onto the drill floor. Down in the subsea supervisor's office, the assistant driller was idly channel-surfing the CCTV system, moving from camera to camera on the rig. He clicked on the camera scanning the drill floor and saw the first spurt of liquid onto the deck. He became alarmed when he saw that the crew was not pulling the drill pipe out of the hole—the normal cause for liquid to appear. There then was a loud noise like a huge pressure relief somewhere and the whole rig jolted, most likely because the massive riser had filled with gas and become buoyant, popping up and hitting the rig. The men in the office did not need to hold a discussion—the subsea supervisor and the first mate ran for the bridge, and the assistant driller headed to the lifeboat deck.

Everyone on the bridge felt the jolt. The spurt of fluid had quickly grown into a cascade, first of seawater, then of more water mixed with drilling mud, reaching the top of the derrick and raining down on the rig, visible on the CCTV screens on the bridge and out the side windows. One of the visiting Transocean managers ran from the bridge to notify the OIM and senior toolpusher.

Down in the senior toolpusher's cabin, the phone had already rung:

> And the person at the other end of the line there was the assistant driller [who] opened up by saying "We have a situation." He said "The well is blown out." He said "We have mud going to the crown."
>
> And I said "Well—" I was just horrified. I said "Do y'all have it shut in?"
>
> He said "[the toolpusher] is shutting it in now." And he said ". . . we need your help." And I'll never forget that.
>
> And I said ". . . I'll be—I'll be right there."[14]

The senior toolpusher pulled on his coveralls and headed across the hall to his office for his boots and hard hat.

Then the gas arrived, its pressure overwhelming the mud-gas separator, and flowed over the rig like a deadly, invisible blanket. The third mate saw gas alarms light up on her display, first one, then another, then all of them. She took a call from the drill floor: "We have a well control issue," then a second one mentioning a "well control situation."[15]

The *Bankston* called, reporting mud coming down on them, and the second mate ordered that vessel to pull a safe distance away from the *Horizon*.

In the first of many notable displays of maritime skill that night, the crew of the *Bankston* detached the hose connecting her to the *Horizon* so it would not fall in the water and foul her propellers, moved quickly to her new station, and launched their fast rescue craft, a two-person speedboat carried on board for man-overboard situations.

Almost immediately three blows were struck in rapid sequence: explosion, power blackout, explosion. The *Horizon* and all aboard her were over The Edge, never to return.

AFTERMATH

One explosion was centered in the area of the drill floor, probably at or near the mud-gas separator, and instantly killed the ten men on the floor and below it in the shaker room, pump room, and mud engineer's office. The explosion destroyed the entire area of cabins and offices, flinging the Transocean manager and the senior toolpusher against the corridor walls and collapsing the OIM's shower around him. Others in that area were buried under the wreckage of partitions and ceilings, and many were hurt.

The second explosion was aft, near one of the engines, and injured the members of the engineering group who were working in their space between the engine banks. The starboard crane operator had remained in his cage forty feet above the steel deck after the first explosion, trying to park his crane according to the emergency drill, and when the power was lost he left it, only to be flung to his death by the second explosion.

The subsea supervisor made it to the confusion on the bridge. Wounded and dazed crew members were filling the room. Four crew members had gone back toward the flames in an ultimately futile attempt to start an emergency generator for fire pumps. The noise from the rushing and burning gases made it necessary to shout. There was uncertainty about what to do.

The subsea supervisor went directly to the emergency disconnect panel and started punching buttons over the objections of the captain. Lights came on, but nothing else happened.[16] They were inextricably attached to the mammoth torch Macondo had become.

The third mate waited no longer. She pushed the button to signal the distress satellite system that covers the oceans of the world, and then picked up the microphone to the emergency radio and made the call heard by those on the *Ramblin' Wreck:* "Mayday. Mayday. Mayday. This is the *Deepwater Horizon.* We are on fire. We have had an explosion and we are abandoning the rig."

No longer a crew, the living made their way as best they could to lifeboats and rafts, helping each other when necessary, and jumping fifty or seventy feet into the burning water when they had to. The *Bankston*'s fast rescue craft shuttled in and out around blazing patches of oil, plucking people from the water and pulling life rafts to safety. They saved every individual who had survived the explosions.[17]

CHAPTER FIFTEEN

The Lessons of Macondo

MACONDO WAS A SYSTEMS FAILURE, and the lessons it teaches are systems lessons. These include different ones for each level of the hierarchy we introduced in Chapter 2 (shown here again in Figure 15.1).

The primary lesson for the hierarchy as a whole is both simple and brutal: undersea blowouts are possible. The pretense that they were not pervaded industry and regulatory thinking before 9:49 p.m. on April 20, 2010. No longer can the question of what to do after a blowout be ignored by industry or dodged through regulatory waivers.

Both the industry and the regulatory regime have made changes, but one major shortcoming that existed before the blowout unaccountably persists: the visible reluctance to incorporate comprehensive and survivable equipment for recording data during safety-critical events on offshore rigs. Such equipment would represent an acknowledgment by those who have installed it that accidents—and even more importantly near-misses—can occur.

If the industry were to install that equipment, it would have to do more than record basic well-control data on pressures, volumes, and flows. It must also record how a crew lines up a well's circulating system and the commands

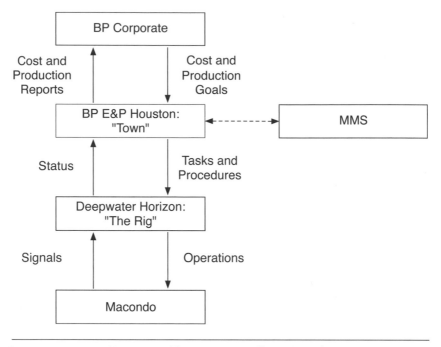

Figure 15.1. The system controlling Macondo.

it sends to the BOP, and provide enough information to allow analysts to reconstruct what the drill crew saw and did. Until survivable recording of such activities is the norm, the forensic capability of offshore drilling will remain firmly in the era of the *Titanic*.

At the corporate level, the lesson for executives and managers is that the perceived obligation to pay stockholder dividends exacts an opportunity cost borne by those who are responsible—directly or indirectly—for keeping the corporation away from The Edge. The result is an essential tension between the part of the corporation tasked with financial performance and the individuals and organizational elements that must ensure safety.

Macondo shows us that if executives resolve that tension significantly in favor of the financial side, by granting high degrees of procedural independence to low-level corporate entities, decentralizing responsibility, relying heavily on outsourcing, and paring down to minimal levels of staffing, as BP did—or engage in other financial engineering steps such as stock buybacks—the resulting opportunity costs can produce a brittle organization incapable of responding effectively to unanticipated events. Financial performance is readily quantified while proximity to The Edge is not, and the seduction of numbers can distort decision making. Any business model for an oil company must adopt the same stance toward hydrocarbons that military doctrine adopts toward the enemy: no one can ignore their presence, and they get a vote in any activity the organization may undertake.

The lesson for project managers is that they play a more important role in keeping an inordinately complex project away from The Edge than conventional wisdom on process safety may indicate. What we call "drilling an offshore well" is not the process of maintaining and operating a pressure vessel that has been previously tested and verified. It is, instead, the design of a vessel that will be constructed in the presence of the pressure it is to safely contain. Both design and construction must proceed with extraordinary care.

What's more, promoting a "safety culture" of methodical wariness is insufficient unless that culture is backed up by an "engineering culture" that includes methodical decision making, contextual review, and management of change. Decisions such as to use foamed cement in an oil-based mud environment, and to set a deep cement plug in seawater instead of mud, require more than a short, two-person conversation. All involved in those decisions and the resulting activities must understand their implications.

Just as important, a corporation must accept that an engineering culture imposes inefficiency in two ways: directly, because of the time employees must devote to those vital efforts, and indirectly, because ensuring that employees at every level take pains with safety-critical decisions slows down other activities. Macondo teaches that those in an oil company who are responsible for allocating resources might save thousands or even millions of dollars by forgoing such activities but spend multiple billions on the other side of The Edge.

The lessons of Macondo for offshore rigs are both technical and managerial. Macondo is a case study in the shortcomings of ad hoc control systems that require a large degree of human communication and coordination to operate without failure. There is no technical reason why a mud logger should not directly know the status of mud transfers off of a rig, why a compliance specialist should not directly know that the flow he is witnessing is occurring while the pumps are stopped, or why the drill crew should not directly know how the circulating system is lined up.

The display of safety-critical information is particularly important in the traditional and persistent management structure of an oil rig, where a drilling team includes employees from different operator, contractor, and oilfield services companies, each with slightly different training and safety regimes. The management lesson of Macondo is identical to that taught by multiple accidents in the aviation industry in the 1920s and 1930s: informal procedures are inadequate when working close to The Edge, and checklists are valuable because they are written in blood.

The final lesson is for all those who have direct or indirect influence on the conduct of offshore drilling. If a corporation in the industry responds to today's extreme financial pressures the way BP did to similar pressures in the 1980s, by adopting a business model dominated by financial concerns and techniques, then the operational side of that corporation will likely evolve as it did for BP: from lean, to brittle, to broken. As Macondo has shown, when part of an oil company breaks, the result can be a very drastic event.

Afterword

Evidence, Reports, and Conventional Wisdom

"LORD, I DON'T KNOW how to tell you guys this, but that was one of the most painful things we could have ever done is stay on location and watch the rig burn. Those guys that were on there were our family. It would be like seeing your children or your brothers or sisters perish in that manner. And that—that put some mental scarring in a lot of people's heads that will never go away."[1]

After rescuing the shocked, dazed, and wounded survivors of explosions that followed the blowout, the *Bankston* remained in the vicinity of the burning rig for ten hours to coordinate the search for the eleven crew members known to be missing. A vessel designed for a crew of fifteen had more than ninety people on board, in bunks, in the lounge and galley, and on decks wet with slippery mud. They were out of cell phone range, and satellite communications were occupied with official business, so they were effectively cut off from the world. Many were emotionally overwhelmed and remained alone with their thoughts. Others engaged in conversation, comparing experiences and trying to make sense of what had happened.

Around 8 a.m. on April 21 the *Bankston* was released from its duties by the Coast Guard and started the voyage to Port Fourchon, Louisiana, stopping at a small rig to board Coast Guard investigators and MMS investigators who had helicoptered there, and lawyers for the owners of the *Bankston*. While under way the investigators selected nine individuals for interviews and passed out accident report forms to the rest of the survivors.

The *Bankston* arrived at Port Fourchon more than twenty-four hours after the explosion. The survivors were required to line up for a urine sample and then were released to medical facilities or local hotels for rest, or allowed to go home. At least one survivor was met at his hotel by Transocean lawyers and interviewed in front of a court reporter before he was allowed to go to bed, and another was interviewed by BP lawyers before he was released.[2]

In the three years between the explosions and the MDL 2179 trial, the survivors were further interviewed by multiple entities: employers, government agencies and commissions, and legal teams. The interviews ranged from wholly informal to those conducted under formal rules of evidence. Some survivors were interviewed as many as five times; the typical number was three. During this period the survivors had access to media accounts and official and unofficial narratives. Multiple crew members also held discussions or took important action several times during the events leading up to the blowout; in every case the accounts of what transpired vary widely.

As a result, the more time that has passed since the event, the more uncertain becomes our primary evidence, as memories are affected by new information, retelling, and the mechanisms whereby wounded psyches attempt to heal themselves. The medications many of the more traumatized crew members were prescribed to deal with anxiety and insomnia introduce other uncertainties. To point all this out is not to be critical of those who have gone through so much, but rather to simply acknowledge their humanity—and to acknowledge as well that an absence of technical evidence has forced all investigators to based their work largely on the frail foundation of human memory.

THE MARINE BOARD OF INVESTIGATION

BP, the oil industry, and the government were not prepared in any substantial way to investigate a disaster of the scope and complexity of the loss of the *Horizon*. In contrast to civil aviation, where accidents are investigated by an organization that is legally chartered to gather knowledge to support changes in technology and procedures, the response to the *Horizon* was uniformly one of improvisation.

The first official investigation was a joint effort between the U.S. Coast Guard and the Department of the Interior, initiated seven days after the blowout. This activity, called the Joint Investigation Team (JIT), began taking testimony on May 7, 2010. The speed with which the JIT was convened is remarkable, considering that the lowest-ranking official with oversight of both organizations is the president of the United States.

The JIT took testimony in the context of a Marine Board of Investigation (MBI), whose rules are set by federal statute and regulation.[3] An MBI is intended as a fact-finding exercise, and as such is granted latitude in its rules of evidence. The federal regulations governing MBIs specify that "no part of a report of a marine casualty investigation . . . including findings of fact, opinions, recommendations, deliberations, or conclusions, shall be admissible as evidence or subject to discovery in any civil or administrative proceedings, other than an administrative proceeding initiated by the United States." This stricture is clearly intended to ensure that investigators use what we have characterized as a scholarly rather than a judicial mindset—a loosely constrained, even-handed search for the truth.

However, in the interest of fairness, the regulations permit those conducting an MBI to name "parties in interest," who have a direct stake in the outcome and are allowed legal representation and right of cross-examination. The *Horizon* MBI named seventeen parties in interest: BP, Transocean, seven other corporations, eight individuals, and the Republic of the Marshall Islands, whose "flag of convenience" was flown by the *Horizon*.

The parties in interest collectively engaged around fifty attorneys, all of whom were aware that the MBI was, in its early sessions, subject to intense media scrutiny as well as a prelude to the inevitable civil trial. The attorneys were therefore determined to both shape the public record and engage in preliminary skirmishes for the inevitable legal actions through cross-examination and objections. In fact, the objections to testimony became so intense that the JIT recessed for a period in July and returned with an extra uniformed Coast Guard attorney and a retired federal judge to help keep order.

The actions of the attorneys for the parties in interest served to shift the MBI away from the scholarly mindset to one that was dominated—albeit indirectly—by a judicial mindset, with significant portions of a limited schedule occupied by partisan cross-examinations focused on assigning blame, and objections focused on avoiding it. Despite this, investigators for the Coast Guard and the Department of the Interior developed an impressive body of knowledge. Their reports were overshadowed by evidence submitted in the massive nonjury trial known as MDL 2179—which by federal regulation essentially had to proceed as if those reports did not exist.

MDL 2179

MDL 2179 occurred in three phases. During the first phase, from February to September 2013, the presiding judge heard arguments regarding whether the plaintiffs acted with ordinary or gross negligence. The second phase, from September to November 2013, litigated the size of the oil spill, to determine the fines the companies would have to pay. The third phase, to establish those monetary judgments, ran for two weeks at the end of January 2015. Only the first phase of the trial produced evidence of interest to our analysis.

Of the *Horizon*'s drilling leadership, only the senior toolpusher testified in open court during MDL 2179. All the remaining leaders either exercised their Fifth Amendment rights, sustained grievous wounds that precluded accurate recollection, or perished in the explosion. (Some of those who refused did testify before the JIT, consented to be interviewed by their employers, or gave accounts to the media.) The senior and junior drilling engineers at BP Town exercised their Fifth Amendment rights and were unavailable for testimony or deposition during MDL 2179.

As we noted in Chapter 1, testimony from those who consented to be deposed or appear in open court during MDL 2179 reflects selection bias inherent in the Federal Rules of Evidence. Most significantly for our concerns, these rules distinguish between a fact witness and an expert witness. A fact witness can give statements only from direct knowledge. An expert witness is entitled to give an opinion. However, to avoid "junk science," expert witnesses in federal cases must meet the *Daubert standard:* they must have credentials in their field and base their opinions on reliable principles and methods. *Fingerspitzengefül* is not admissible, as shown by the fate of this question asked by a plaintiffs' attorney: "Do you have—based on what you knew that night, what you've learned since as a senior toolpusher on this rig, have you formulated your own opinion, based on your education, your training and your experience as to exactly how this well could have been allowed to blow out the way it did the night of April 20, 2010?"[4]

A defense attorney objected on the grounds that the supervisor was not an expert witness in the eyes of the law, and the objection was sustained. And so the world has been deprived of the possible insights of an individual who had thirty-three years of offshore experience, had been on the rig since it was built, and directly supervised the drill crew. The depositions of other witnesses were likewise redacted by the court based on the Daubert standard and other rules of evidence such as the prohibition against hearsay, so selection bias extends to this material as well.

On September 4, 2014, the presiding judge issued a ruling on the trial's first phase.[5] He found that BP was guilty of gross negligence, and that Transocean and Halliburton were guilty of ordinary negligence. He assigned BP 67 percent responsibility for the event, Transocean 30 percent, and Halliburton 3 percent. The remaining defendants, all subcontractors of BP, were released from liability before or during the trial. Halliburton settled with the plaintiffs on the eve of the judgment and was also released from liability.

The presiding judge issued "findings of fact" in support of his ruling: his assessment of the causes of the disaster. He limited these findings almost exclusively to the events of the last day. He ruled that the cause of the blowout was a casing breach, and therefore that the quality of the cement job was irrelevant. He blamed the well site leaders for the misinterpretation of the negative test, and exonerated the Transocean crew for failing to recognize the symptoms of a kick.

He also ruled that the BOP failure was due to poor maintenance on the part of Transocean, and that the crew had acted "appropriately and bravely" after the explosions. However, he concluded his findings by ruling that neither Transocean's maintenance practices for the rig as a whole or BP's process safety management system were causal factors. Any differences between his findings and our narrative can be attributed to the fact that he was, by law, permitted to consider only evidence allowed by the rules, whereas we have had the freedom to consider evidence from all available sources.

On January 15, 2015, the presiding judge issued his ruling on the second phase. He decided that BP, for purposes of computing fines, had spilled 4 million barrels of oil into the Gulf. The company received "credit" for 810,000 barrels collected from the ocean surface after the spill. The final phase ended with a negotiated settlement in which BP agreed in principle to pay $18.7 billion to the plaintiffs over a period of eighteen years.

LIMITS OF THE EVIDENCE

Early on it became clear that the *Horizon* disaster had left behind very little hard evidence. There was no wreckage to provide "witness marks" on the configuration of critical elements of the rig's circulating system until a government team retrieved the BOP five months after the event. In fact, the BOP and a section of the riser are the only parts of the *Horizon* that were recovered.

Even this activity was surrounded by uncertainty. In May, crews used an ROV to remove one of the crucial control pods and bring it to the surface. Technicians from Cameron International, the BOP's manufacturer, then essentially refurbished and tested this control pod over a ten-day period. It was then lowered to the BOP and reinstalled as part of the ultimately aborted "top kill" attempt to stem the flow of oil from the well. This effort would have entailed pumping heavy drilling mud down the choke and kill lines of the BOP.[6] In July the other control pod was brought to the surface and subjected to a similar upgrade, test, and installation in support of the ultimately successful "capping stack" operation, which involved placing a small version of a BOP on top of the one on the sea floor.

Technicians performing this work were careful to log what they had done, and handed over any parts they removed to the custody of the FBI, but they were engaged in a time-critical remediation exercise, not a forensic examination. Their actions, critical to the killing of the well, insert an unavoidable element of doubt into the results of later testing of the control pods.

After the well was plugged in September, the *Q-4000*, a specialized well-intervention vessel, salvaged the BOP's 400-ton stack of valves and placed it on a barge for transfer to a NASA test facility under FBI supervision. There the stack was subjected to six months of testing and analysis by Det Norske Veritas (DNV), an international inspection, safety, and engineering firm. The DNV tests primarily entailed physical examination of the rams and annular preventers of the BOP and the drill pipe found trapped inside it. The company also tested the control pods by restoring them to what was believed to be their condition at the time of the blowout. It did so by replacing upgraded parts with those removed during the May and July 2010 remediation. This testing focused on various emergency disconnect mechanisms and did not examine controls for the lower kill line valve that was potentially a cause of the crew's misjudgment of the second negative test.[7]

The only other technical evidence that survived the explosion—the kind of evidence provided by the "black box" in an aircraft accident—existed in the form of data from the Sperry Sun system. Event data recorders—the formal name for black boxes—have been installed in aircraft since the late 1950s, heavy highway vehicles since 1998, Formula One racing cars since 1997, passenger automobiles since 1994, and seagoing vessels since 2002.[8] The *Horizon* was equipped with the marine version, called a *voyage data recorder,* but it preserved information of interest only to the marine side, such as the position and speed of the vessel. This recorder was not recovered from the wreck.

The Sperry Sun Data

The system that everyone associated with the *Horizon* disaster calls "Sperry Sun" was officially known as *INSITE Anywhere,* provided by Halliburton, which had acquired Sperry Sun in the early 1990s.[9] The system was marketed at the time as a tool for improving drilling efficiency, not as a specialized forensic recording device:

> If you have Internet access and a Web browser, you can access well logs from anywhere in the world using the INSITE Anywhere™ service. As data moves from your logging tools to a secure web site operated by Halliburton, your asset team can review the results in real time and make collaborative decisions on the best avenues to pursue.
>
> INSITE Anywhere is all about making the most efficient use of your time and budget. The system even allows you to participate in multiple wellsite operations from a single location. Flexibility and functionality are the watchwords of the system, and displays of everything from log plots to pressure tests and samples can be configured to individual preferences.[10]

The ability to remotely store and view data was an extension to the facilities installed by Sperry, the company that provided the mud logger. During active drilling, mud loggers have a variety of duties. During the production tail phase of Macondo, the mud logger provided a second set of eyes monitoring the basic well-control parameters of pit volume, flow rates coming out of the well, and pressures at various locations in the circulating system. These data were captured by a mix of sensors, some provided by Sperry and some by NOV as part of the HiTec system, which displayed the status of the circulating system.[11]

Besides being transmitted ashore, Sperry Sun data were displayed in the mud logger's office and on a single screen to the left of the A-Chair in the drill shack. In the latter location, the Sperry Sun display was largely redundant to the HiTec display directly in front of the chair. Drillers and mud loggers could select a digital format, but the most commonly used view duplicated an old-fashioned paper strip chart that scrolled by as time passed, as shown in Figure Aft.1.[12] (Color-coding of the curves—not shown here—slightly improves the readability of the display.)

Beyond the basic data of pressures, flows, and pit volume, the Sperry Sun equipment also displayed and transmitted the amount of hydrocarbon gas that was returning with drilling fluids from the well, and information on the draw works, such as how much weight the works were lifting and how high.

Many analyses of the blowout, starting with BP's own report, criticized the crew's actions based on inferences drawn from the Sperry Sun data. Such inferences suffer from two forms of uncertainty: the unknown accuracy of the Sperry Sun data, and a lack of knowledge regarding exactly what crew members were looking at on their screens when they were attempting to determine what was happening down in the well.

The accuracy of the Sperry Sun data available to us can be influenced by two factors: the accuracy of the sensors, and the way the data are "packaged" for transmission ashore. Readings from the basic sensors measuring pit volume, flow, and pressure had inherent uncertainties. All the sensors were also years overdue for recalibration, and, as we noted in Chapter 11, some were distrusted by the crew.

The conversion from continuous (analog) sensor readings to digital format for transmission to BP Town could also undermine the accuracy of the Sperry Sun data. This process involves sampling sensor readings at fixed intervals. Typical sampling algorithms can induce small errors through scaling—"trimming" the reading to fit a specific format, such the "ASCII"[13] format used to transmit the information to onshore receivers. The Sperry Sun system

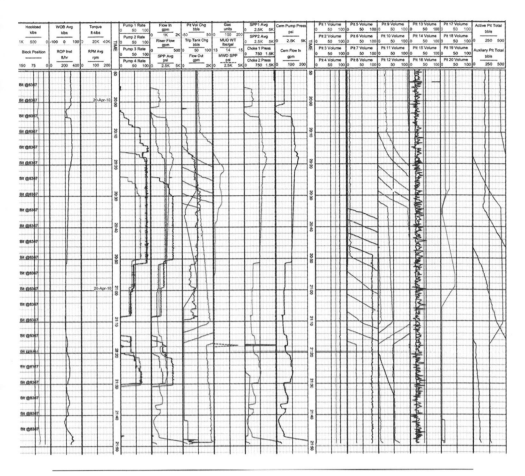

Figure Aft.1. The Sperry Sun curves for the final hour.

also uses an unknown smoothing and averaging process, along with unit conversions and sensor calibrations, to convert the transmitted data to the curves used by the majority of forensic analysts. This process not only can introduce absolute errors but also can make abrupt changes in sensor readings appear gradual. None of the teams investigating the disaster (see below) examined this conversion process, and none except a laboratory hired by Transocean had access to data in the ASCII format.[14]

Many post-blowout analyses of the Sperry data also exaggerate the vertical axis of significant curves, such as the pressure in the well just before the explosion. These exaggerated recreations suggest that the display might have revealed an obvious change in such pressure, when in fact a driller may have set the display to a more compressed scale, or to show pit volume or flow rather than pressure.

NOV Refuses to Cooperate

Investigators could gain more useful insights by conducting an exercise similar to those used to analyze aircraft accidents, in which final flight patterns are "flown" on simulators to shed light on crew behavior. Such an exercise would first require estimating how the drillers who perished typically configured the HiTec system, based on information from those who had worked with them. Investigators could then feed Sperry Sun data—formatted to look like the values from relevant sensors—to a HiTec system.

The chief counsel of the National Commission on the BP *Deepwater Horizon* Oil Spill and Offshore Drilling, whose report we describe below, did in fact attempt to conduct such an exercise, and asked NOV to provide a HiTec drill-control system for that purpose. However, NOV denied the request on the grounds that "manufacturing guesses as to what was displayed on the rig's computers runs a serious risk of producing a misleading picture of what actually happened."[15] NOV further contended, "It is impossible to say what channels or parameters, time scales or pages the driller had displayed at the time of the accident."[16] The company's attitude is quite possibly unique in the annals of official investigations of major accidents. As a result, we do not have even a plausible set of "what if" scenarios for evaluating the human factors associated with the last minutes before the *Horizon* blowout.[17]

INFLUENTIAL REPORTS

Several agencies and organizations—along with BP and Transocean—issued reports on the disaster before the civil trial. All these reports reflect selection bias stemming from the sponsor's focus and agenda. In this section we cover those with the most visibility and influence on both professional and public views of the event, and whose conclusions—whether justified by the material that came to light in MDL 2179 or not—have attained the status of conventional wisdom.

Reports of the Joint Investigation Team

The JIT released its findings in two parts. The first, the *Coast Guard Report of Investigation,* appeared in April 2011, followed five months later by the report of the Department of the Interior. The split reflects the two cultures of the agencies, whose differences work against the analysis, design, and management of rigs like the *Horizon* as a single system. The marine culture tends to view such a rig as a seagoing vessel that happens to carry drilling equipment. The drilling culture views the rig as a drilling system that just happens to float on the ocean.

The Coast Guard report[18] covers the explosion, fire, and evacuation of the rig. It places the accident in the context of the elaborate structure of regulations and safety codes that govern maritime activities, and points to discrepancies between the state of the rig and the requirements of the International Safety Management Code. However, it does not note these discrepancies as causal factors. The report is highly critical of the maintenance and management practices of Transocean, as well as the regulatory and inspection practices of the Republic of the Marshall Islands, including the dual command structure the Republic permitted to exist on the rig.

The Department of the Interior report[19] covers the time between the completion of drilling and the explosion. The panel that produced it relied on outside experts to interpret the Sperry Sun curves. This was the first investigation that had access to DNV's forensic evaluation of the BOP. The value of that evaluation, while significant, was diluted by the fact that the electro-mechanical control pods—which played a key role in the failure of the BOP—had been retrieved, refurbished, and reinstalled and then retrieved a second time with the rest of the BOP before being tested.

The Department of the Interior report includes some of the earliest descriptions of the lack of engineering discipline at BP Town. It also notes—but does not emphasize—the subsequently validated conclusions of its consultant that the well was losing mud into the formation during the second displacement to seawater until 9 p.m., when it started to kick.

The BP and Transocean Reports

While the MBI was taking testimony, BP and Transocean began their own internal investigations. Each company interviewed survivors, and each

allowed the other to interview its employees. Both BP and Transocean conducted their investigations informally: interviewers took cryptic handwritten
notes rather than consulting verbatim transcriptions of recorded interviews.
Both corporations also engaged outside consultants to conduct specialized
investigations, such as simulating the flow of hydrocarbon gas over the rig
after the blowout.

The BP *Accident Investigation Report*[20] was the first comprehensive study
of the accident made available to the public and media. It was issued in September 2010, roughly coincidental with the plugging of the well. As we noted
in Chapter 1, it adopted a judicial mindset and was defensive in nature,
using the "Swiss cheese" model to ascribe causality to everyone but BP, and
omitting discussion of systemic issues such as the company's management
and engineering culture, although it does criticize some of the actions of BP
Town. Nevertheless, the report offers no evidence of the kind of deep corporate soul-searching that Exxon Mobil underwent in the aftermath of the
Exxon Valdez spill.[21]

The report uses exaggerated versions of the Sperry Sun curves to support
the thesis that the drill crew was inattentive, and it does not address the issue
of narrow drilling margins and lost returns, during either active drilling or
the final displacements. Whether by accident or design, the report does not
explain the role of the BOP in well control as we did in Chapter 3, leaving
lay readers with the impression that the BOP is primarily an autonomous or
semiautonomous safety device. This impression—reinforced in remarks by
BP's CEO—persists to this day, perpetuated in media such as the *New York
Times*.[22]

Transocean released its report[23] in June 2011, in between the two JIT reports. The Transocean report underscores the company's position that its
employees were simply technicians who followed the orders of BP managers,
who ordered them to conduct risky activities and failed to convey the nature of those risks to the Transocean crew.

The report attempts to illustrate what the drill crew saw on its NOV
displays by artistic means, rather than by running data through an actual
NOV system. As noted, this approach can be misleading, because there is
no guarantee that the smoothing method used by Sperry Sun would yield
the same curves as the method used by the NOV system. Any difference
would be significant in cases of sharp changes in pressures in the well. Static
representations on paper of NOV displays also do not convey the rate at

which such readings are changing—an essential factor in understanding possible reactions by the drill crew.

Report of the National Commission

President Obama established the National Commission on the BP *Deepwater Horizon* Oil Spill and Offshore Drilling in May 2010. The commission's report,[24] issued in January 2011, covers the period between completion of the drilling phase and the stopping of the well flow in September 2010. The report focuses on the impact of the spill on the environment and the economy of the Gulf.

The commission was highly critical of the major players in both government and industry, stating that "efforts to expand regulatory oversight, tighten safety requirements, and provide funding to equip regulators with the resources, personnel, and training needed to be effective were either overtly resisted or not supported by industry, members of Congress, and several administrations."

The report lists major decisions by BP that saved time or money while adding to risk and management shortcomings. However, it does not examine systemic problems stemming from BP's overall business structure. The report's technical analysis—apparently by an outside expert—is based largely on the Sperry Sun curves. The technical conclusions echo those of the BP report and do not include any description of lost returns during the final displacement.

Report of the Chief Counsel to the National Commission

The chief counsel to the National Commission released a report a month after that of the parent organization.[25] His report describes events from the end of the drilling phase to the explosion, and considers well design, what should have happened with the well, and what the evidence suggests actually did happen.

The report draws its conclusions from MBI evidence, internal BP and Transocean documents, and interviews by the chief counsel's staff. About a quarter of its references cite confidential sources. (A bill awarding the chief counsel subpoena powers passed the House of Representatives but died in the Senate,[26] so he had to limit his investigation to evidence volunteered by corporations and individuals.)

The chief counsel restricted his analysis of management shortcomings to BP Town, overlooking the impact of BP's corporate structure. The management-related discussions appear in disjointed sections, obscuring the timeline of decisions regarding Macondo and the pressure to complete the well that was clearly manifest in the actions of BP Town. The well-control discussion depreciates the notion—later validated—of lost returns during the final displacements, and misconstrues the contents of the InTuition Energy paper. The report also relegates the important topic of Kaskida to an appendix (see the next section).

KASKIDA AND THE RUSH TO FINISH

Many of these reports promote the conventional wisdom that the extreme haste during the *Horizon*'s last few days reflected the "time is money" culture of the oil business in general and BP in particular. For example, the Chief Counsel's Report categorizes seven of nine decisions that increased risk under the heading "Saved Time":[27] "Given the many decisions that increased risk but saved time and money, it is a reasonable inference that cost and time overruns had an effect, conscious or unconscious, on decision making."[28]

This approach conveys the impression that the event was like a biker who pressed a bit too hard at a bend and was unlucky enough to go over The Edge. Our assessment is that it was more like a motorcyclist tearing down the freeway, splitting lanes and weaving between cars at top speed.

The pressure to complete Macondo manifested itself in two ways: go fever and the pace of the final activities on the *Horizon*. The senior mud logger on duty at the time noted the highly unusual pace in his testimony:

> Q. Now, is it true, is it not, [. . .] that during a normal displacement—
>   you've been through this kind of procedure before, have you not?
> A. Yes, I have.
> Q. During the normal way that this was done, everything would be at a
>   standstill, correct?
> A. Yes, sir.
> Q. There would be no rig activity, correct?
> A. Correct.
> Q. Nine out of ten times, BP would cease all other activity because this is a
>   safety-critical activity, correct?

A. Correct.

Q. And in your 18 years of working offshore, you have never seen all of these simultaneous activities going on during displacement; isn't that true?

A. It's never been like this before.[29]

What the mud logger described is known in drilling as simultaneous operations, or *simops*. Those activities included running the system as an "open loop" after the sheen test, thereby bypassing the Sperry flow sensor and the CCTV observation of flow; emptying the trip tank multiple times; emptying the bins containing cuttings extracted from the returning mud; transferring that mud between pits to accommodate cleaning crews working on them; and running the gantry crane on the aft deck back and forth during preparations for setting the lockdown sleeve.[30]

BP management recognized simops as dangerous because they could distract the crew and also insert "noise" into the signals provided by the circulating system, and BP's governing documents mandate a risk analysis before engaging in them.[31] There is no evidence that anyone on the *Horizon* performed such an analysis.

However, other crew members noticed the unusual pace of activity, as described by the chief electronics technician:

Q. Around this time on April 20th, was there pressure to speed up operations on the well?

A. Always.

Q. Okay. What do you mean by—by that?

A. Just a general consensus around the rig that the—the bonus was, you know—the well bonus[32] was out the window. We were way far behind. We had been stuck real bad in the—in the middle of this well and lost, you know, several million dollars in fluid and equipment. You know, we had already started conducting training for the next well that we were going to, which was Kaskida. Our Sunday Safety Meetings were already starting to focus on that. So it—it just seemed like, you know, they were in a real big hurry to get out of here and get to the next one.

Q. You said "general consensus." Is that—you referring to all the workers on the rig?

A. The people that I spoke with, yes. I had interaction with—with, like I said, everyone, from the Cook to the Maintenance Supervisor and, you know, anyone in between. H'm, a Chief Electronics Technician doesn't have a lot of pull on a rig, but I've got a lot more than most, so people would sometimes voice these things to me.

Q. Where did this pressure . . . originate from, from your perspective?

A. Just how fast they were drilling, how fast they were moving, how hard they were pushing to—to get this thing completed and get on to bigger and better things.[33]

In a later interview with BP internal investigators, the substitute WSL noted: "They decided we could do the displacement and negative test together—don't know why—maybe trying to save time. At the end of the well sometimes they think about speeding up."[34]

There was indeed a relationship between time and cost, but like most things about Macondo, it was anything but straightforward.

Cost Control at BP

BP managers recognized two broad "colors of money" in their decision making. The first was the day rate BP paid for the *Marianas* and the *Horizon*. The company was obligated to pay this rate—roughly a million dollars a day—from the time it signed leases on the rigs until the end of 2009 in the case of the *Marianas,* and the end of 2010 in the case of the *Horizon.*

Cost pressure in such contractual relationships becomes utilization pressure: the urge to get the most out of the rigs. Utilization pressure leads to "drilling like a bat out of hell" and other steps to reduce idle time: "In an interview with the Chief Counsel's team, [the operations manager] shared that he was always thinking about how to drill wells faster."[35]

Utilization pressure also has a negative impact on the state of repair on the rigs, as it leads to a general unwillingness to stop operations to fix things. All oil companies feel such pressure, but BP's overall decision to put "tremendous pressure on costs" exacerbated it.

The second "color of money" was discretionary expenditures: money that could be spent or not, depending on the decisions of managers. Discretionary expenditures were the target of BP's Make Every Dollar Count campaign of 2009.[36] When applied to understaffed, brittle organizations such as BP, a squeeze on discretionary expenditures affects the operation as a whole as well as individual projects.

Such a squeeze was the only apparent reason for dangerous decisions made at Macondo, such as the choice of an inappropriate cement mix because it was already on the rig; using for a spacer leftover material that no one had ever used for that purpose before, and which had an unknown effect on drillers' ability to "listen to the well"; and the decision to use drill pipe already on the rig to weight the lockdown sleeve—a decision that adversely influenced other elements of the plan.

The Phantom Cost Overrun

In October 2010 a *Washington Post* article asserted that Macondo had sustained a cost overrun of $58 million on an initial budget of $98 million— an overrun of 61 percent.[37] The chief counsel and the presiding judge for MDL 2179 reiterated this assertion,[38] and accounts of the disaster have repeated it ever since.[39] This assertion strongly implies that the *Horizon* crew's dangerous haste in conducting activities related to the production tail was motivated solely or primarily by a desire to minimize that overrun. However, BP managers have consistently contended that the budget for Macondo did not put pressure on the team to complete the well. Evaluating these conflicting accounts requires looking deeper into the project's fiscal history.

Two classes of documents governed expenditures at Macondo. The first and highest-level was an *Executive Financial Memorandum*. The EFM was approved by the chief of exploration and production in London[40]—essentially the head of BP's "upstream" operations, which include locating and extracting oil and gas. ("Downstream" operations include refining and marketing.) This individual was a member of the BP board and held a position just below the CEO.

An EFM authorization includes a target amount and a not-to-exceed amount. If operations at a well are in danger of exceeding the latter, project managers must obtain a new authorization. BP's practice evidently was to drill a well to a stated milestone such as target depth, and to add funds to the target amount as needed along the way. One BP budgetary official said he was not aware of any case where the company had abandoned a well for fiscal reasons.[41]

The second class of document is *Authorizations for Expenditure* (AFE), which release funds budgeted by an EFM. These documents control the budget of a well project and must be reauthorized within EFM limits if the project exceeds the budget. AFEs also include an anticipated completion date. The vice president of exploration for the Gulf of Mexico—two levels of management below that required for an EFM—authorized AFEs for Macondo.

The relationship between the two funding documents was not rigid: the vice president signed an AFE for Macondo before the governing EFM was signed, and managers spent funds in excess of the AFE limit before obtaining supplemental authorization. These instances are consistent with the BP budgetary official's testimony that funds for well projects were controlled, but not so rigidly that they might be canceled if they exceeded limits.

The somewhat convoluted history of these documents for Macondo suggests that expenditures on the well attracted a relatively low level of concern from top management. The alacrity and lack of complaint with which partners in the Macondo project authorized funds for it—and the absence of review meetings and replanning that occurs in other engineering environments when a project exceeds expected costs—reinforces this notion:

June 18, 2009: The vice president of exploration for the Gulf of Mexico signed the first AFE for Macondo, with a budget of $124 million. The authorization included $21.4 million to cover the day rate on the *Deepwater Marianas* if she was deliberately idled ("stacked") to avoid hurricanes. At this point BP was underwriting all expenses.[42]

August 28, 2009: Anadarko, MOEX, and BP jointly signed a second AFE in which they agreed to share expenses incurred at the well. The partners reduced the authorization to $96.1 million because it no longer included an allowance for idle time. BP's fiscal risk dropped to $64 million.[43]

September 28, 2009: The first EFM was authorized, with a target of $96.1 million and a not-to-exceed of $139.5 million.[44]

November–December 2009: BP executed a cost-sharing agreement with Anadarko and MOEX that involved profit sharing.[45]

January 12, 2010: According to internal email from BP financial operations, gross spending on Macondo totaled $71 million to this point, of which BP's share was $40.2 million.[46]

January 27, 2010: The three partners signed the first supplementary AFE, adding $27.9 million to the gross budget and increasing BP's contribution by $20.9 million, to cover funds spent on the *Marianas* while she was in dry dock.[47] As we've noted, BP was obligated to pay the day rate on *Marianas* until the end of 2009 no matter where she was or what she was doing, so this "overrun" simply shifted costs among projects.

March 22, 2010: The partners approved a second supplementary AFE that added $27 million to the gross budget, which then totaled $151 million. BP's share again rose by $20 million. This was the final AFE for the exploratory phase of Macondo.[48]

March 24, 2010: A supplemental EFM was signed, raising target gross expenditures to $151 million and the not-to-exceed to $166 million. BP's share of this budget was $99 million.[49]

April 14, 2010: The partners approved an AFE for installing the production casing, cementing it in place, and installing the lockdown sleeve for $3.5 million, of which $2.3 million was allocated to BP.[50]

By conservative estimate, $15 million remained in the authorized budget for Macondo when it blew out. What's more, BP was obligated to pay the *Horizon*'s day rate until the end of 2010 no matter which well she was working on. Thus the cost overrun for the project was more on the order of $20 million. It is highly unlikely that the dangerous level of activity during the *Horizon*'s final days was prompted by a management trying to save a small percentage of an already authorized budget. We need to consider other sources of scheduling pressure.

A Game of Musical Rigs

To do so we must reconstruct the problem that BP faced in scheduling rigs in the Gulf of Mexico. The drilling rigs the company had contracted had to complete at least three tasks in mid-2009: plugging the Nile well, drilling an appraisal well[51] at Kaskida, and drilling Macondo. The Nile well had to be plugged and abandoned before July 2010, under a regulation that companies had to render depleted wells safe within one year of the end of production.[52] The Kaskida situation was potentially more serious. Kaskida was—and is—one of the largest prospects in the Gulf, with estimated reserves in excess of 3 billion barrels of oil-equivalent, of which BP owns 70 percent.[53] BP could have lost its lease for that mammoth reserve by missing a regulatory deadline to start work at the site by May 15.[54]

In mid-2009 BP had under contract two rigs that were finishing up wells and would be free to undertake these tasks. The *Deepwater Marianas* was completing one well called Na Kika H-2. The *Marianas* lease was scheduled to expire in early December 2009, at which time she was to leave BP and go to work for an operator called ENI.[55] The *Deepwater Horizon* was drilling a well called Kodiak, and her lease was not scheduled to expire until September 2013.[56] We can produce a speculative but plausible scenario regarding how BP planned to assign these rigs.

The *Horizon* was no stranger to the Kaskida prospect, having drilled the Tiber exploratory well there during the first half of 2009, which then held the world record for depth at 36,100 feet. The rig was also more suited to drilling an appraisal well at Kaskida by virtue of being dynamically positioned and specializing in exploratory wells.

The *Marianas,* less suited for Macondo owing to her focus on production drilling, was probably a better choice to deal with the first subsea plug and abandonment of a once-producing BP well in the Gulf.[57] A logical schedule, given what was known at the time, would have been for the *Marianas* to drill Macondo between August and November 2009, and then be towed 38 nautical miles to Nile, where she would perform the thirty-day plugging operation before her release to ENI. *Horizon* would finish up Kodiak, undergo several weeks of much-needed maintenance, and then proceed under her own power the 220 nautical miles southwest to Kaskida.

Events from mid-2009 to April 2010 conspired to disrupt these plans. First, the *Marianas* was delayed by BOP problems and a bad cement job at Na Kika.[58] Then she was put out of action by Hurricane Ida, and spent the rest of her lease period in dry dock before leaving BP's service. The *Horizon* was also delayed in starting at Macondo, and once there its schedule slipped thirty more days because of well-control problems. Of even more concern, Dril-Quip—which BP had contracted to produce a new model wellhead to handle the greater depths and higher pressures at Macondo—had run into difficulties and pushed back the delivery date to May 1, just fifteen days before the regulatory deadline.[59]

Planning for Kaskida had started even before the *Horizon* had completed exploratory drilling at Macondo, with a "crew engagement meeting"—a kind of kickoff meeting for new wells—at the end of March.[60] The *Horizon*'s Sunday safety meetings were beginning to focus on Kaskida. By January 2010 BP had appointed a manager under the vice president for exploration and production to oversee the Kaskida appraisal, and had formed a group to oversee the conversion of Kaskida to a production well.

By April 9 BP managers had two choices regarding the scheduling of the *Horizon.* They could send her off to Kaskida to perform what was essentially busy work to meet the regulatory deadline for Kaskida, or they could attempt to squeeze the Nile task into her schedule. The first option meant incurring substantial idle time, and the second option put pressure on her crew to complete Macondo. BP managers chose the second—hedging their bet by submitting a request for relief from the deadline to MMS. Such requests, in the words of the BP negotiator who prepared it, were "not slam-dunks," "particularly now-a-days."[61] BP submitted the formal request on April 16,[62] and had received no decision before the *Horizon* sank.

We do not know who chose the second option, which factors influenced that decision, or whether anyone objected. Nor do we know the likely reaction of MMS to BP's request for relief from the deadline. Our uncertainty reflects the selection bias that arose from the judicial mindset that dominated so many of the investigations. Although the choice of the second option arguably sealed the fate of the *Horizon* and eleven of her crew, it remains largely unexamined because none of the parties in the lawsuit apparently thought that exploring it would enable them to either place or avoid blame.

Denials

After the disaster BP executives denied that they had applied pressure to move the *Horizon* off Macondo, as exemplified by this careful exchange between the lead attorney for BP and the vice president for drilling and completion:

Q. Were you the person responsible for rig scheduling?

A. Yes.

Q. Okay. At any time, did you ever express a concern to either your direct reports or to anybody in the D&C organization that—that they should speed up operations at the Macondo Well because of concerns regarding the rig schedule and Kaskida?

A. No.

Q. Did anybody from Transocean on the evening of April 20th, while you were with them—

A. M-h'm.

Q. —for that day express to you any concerns that they felt BP was pressuring them to speed up the pace at which they were operating at the Macondo Well?

A. No.

> *Q.* Did you ever feel that the Kaskida lease was at risk because you didn't
>   have a different rig to satisfy the requirements of meeting the MMS
>   application if the MMS did not approve the extension?
> *A.* No.[63]

Similarly, the chief operating officer for exploration and production
testified:

> *Q.* In your [COO] review, did anybody communicate to you that they had
>   pressure out at the Macondo site because they were behind time and
>   they might not get to Kaskida on time?
> *A.* I don't recall that explicitly.[64]

This answer echoes the reply of the fired vice president (as noted in
Chapter 2) who, when asked if BP cut costs at the expense of safety, paused
and answered, "Not explicitly." That individual described the alternative in
a speech he gave to a professional society five months after the disaster: "The
"elephant in the room" is all the mixed or unintended messages we send the
crews when we are behind schedule, over cost, or behind on production. If
we don't clearly keep personal and process safety as an unyielding value in
our words and more importantly visible behaviors and decisions, we ultimately will not withstand the risk or test of time, and we will certainly suffer
a fatality or major incident."[65]

In his report, the chief counsel relegated the Nile and Kaskida issues to
an appendix, and dismissed their significance with these words: "Though
BP's decisions at Macondo appear to have been biased in favor of saving time
and money, the rig's next wells do not appear to have been an important
contributing factor. BP followed the rig's schedule closely and, when necessary, took action to relieve the pressure of regulatory deadlines."[66]

This statement is both true and misleading. It is the essence of an emergent event that individual factors do not appear significant when taken out
of context. Rather, the way they interact and combine causes the event. The
individual factors were the continual background messages to those working
on Macondo conveying the corporate attitude toward cost and delay, the
natural desire of *Horizon* crew members to put a difficult well behind them,
and the equally natural desire of a largely new management team to focus
on a new project. Macondo was old news and bad news, to be wrapped up

as quickly and with as little effort as possible. Work on Kaskida had already begun, drawing the attention and diverting the effort of an already brittle and understaffed organization in Town.

No explicit pressure from BP upper management was therefore required, and it is plausible that none was applied. What was required for survival was explicit pressure on crew members to slow down, determine how close they were to The Edge, and take steps to move away from it—pressure that never came.

Appendix

Technical Notes

These notes expand on technical topics introduced in the main narrative. They support assertions in that narrative while also providing material for those who wish to learn more about the technology of offshore drilling.

Technical Note 1 presents the formulas used in primary well-control calculations, first described in Chapter 3 and referred to throughout the rest of the narrative.

Technical Note 2 describes the mechanics of the blowout preventer, whose role in secondary well control we also introduced in Chapter 3. This note also summarizes what other investigators have discovered or asserted about the BOP's failure to automatically shut in the well.

Technical Note 3 explains the concept of drilling margin, and how the need to maintain it determines when a crew must run new intervals of casing in a well.

Technical Note 4 covers the mechanisms of the float collar and wiper plugs, which we described in purely functional terms in Chapter 7.

Technical Note 5 provides an analysis of the Sperry Sun data that supports the assertion, by ourselves and others, that Macondo was losing returns during the first and second displacements to seawater, as described in Chapter 13.

Technical Note 6 provides more detail on the physical configuration of the *Horizon*'s circulating system, and shows how that layout contributed to the confusion and missed communications during the last hours depicted in Chapter 14.

TECHNICAL NOTE 1: WELL-CONTROL CALCULATIONS

The predictive calculations at the heart of primary well control are based on the science of liquids at rest, or *hydrostatics*. A fundamental principle of hydrostatics is that the pressure of a body of liquid at a given depth depends on that depth, the density of the liquid, and nothing else. Thus, a pressure gauge lowered to a depth of fifty feet in saltwater will read 36.3 pounds per square inch (psi), whether the reading is taken in a saltwater tank at an aquarium or in the middle of the Pacific Ocean.

The main liquid of interest to drillers is drilling mud, although other fluids are used for specific tasks, such as separating drilling mud from seawater during a displacement. Originally mud was just that: local dirt mixed with water and oil. Today drilling mud is a sophisticated mixture of a base fluid, such as synthetic oil, and a powdered mineral, such as barite.

BP, Transocean, and the MMS measured the density of drilling fluids in pounds per gallon (ppg).[1] Drilling mud can vary from 8 ppg (freshwater) to 22 ppg. The depth of interest is that of a vertical (or nearly vertical) hole in the ground.

The Magic Number 0.052

The most basic formula of well control provides the pressure at any depth of hole:

> *Multiply depth in feet × mud weight in ppg × 0.052 to get the pressure at that depth in psi.*[2]

This defines the amount of downward "push" that resists the upward "shove" of pore pressure.

A little bit of algebra yields another formula of fundamental importance to well control:

Multiply the depth by 0.052 and divide the result into the desired pressure to get the mud weight that will generate that pressure.

To see how that formula works, consider a well that has been drilled to 1,000 feet—a zone that generates a pore pressure, or "shove," of 624 psi. To keep the well under control, we need a "push" of 624 psi at 1,000 feet. So we calculate:

$$1,000 \times 0.052 = 52$$

and then

$$624/52 = 12$$

So the desired mud weight is 12 ppg.[3]

These two formulas are also the basis for the way drillers express pressure at a depth. Instead of an absolute value, such as psi, they use the weight of mud required to generate that pressure at the depth of interest.

In actual practice, the drill crew would most likely be given the geologist's initial assessment of our hypothetical well in terms of depth and ppg. The zone in the example above would not be called a "624-psi zone," but rather a "12.0-ppg zone at 1,000 feet." Drillers would then use the first formula to derive absolute pressures when needed.

Figure TN1.1 also shows the different absolute pressures associated with the value 12 ppg at different depths.

The U-Tube Effect

A pipe inside a body of fluid forms a structure called a U-tube—so-named because it has the hydrostatic characteristics of a U-shaped tube, where one arm is the pipe and the other is the annulus. This, in turn, leads to the phenomenon known to drill crews as *U-tubing*. This occurs when one arm of a tube is sealed and filled with a light fluid, and the other is left open and filled with a heavier one. The heavier fluid will apply pressure to the lighter one,

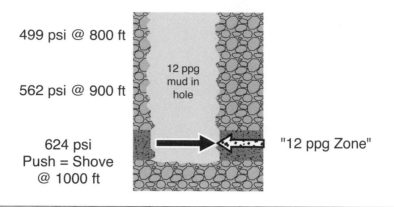

499 psi @ 800 ft

562 psi @ 900 ft

624 psi
Push = Shove
@ 1000 ft

12 ppg
mud in
hole

"12 ppg Zone"

Figure TN1.1. Basic well-control calculation.

depending on their relative heights and fluid weights, and, like all such hydrostatics, independent of the diameters of the tubes.

During well construction, U-tubing occurs most often when lighter fluid such as seawater displaces heavier fluid such as mud, as was the case during the last hours of Macondo. Crews can use the basic well-control formulas to calculate the amount of pressure on the less dense fluid, which will then tell them what pressure to expect at the surface in a U-tube situation (Figure TN1.2).

Consider a 2,000-foot well filled with 13.5-ppg mud. The crew runs 1,500 feet of pipe into the well, and pumps in 8.6-ppg seawater. When the seawater reaches the bottom of the pipe (having pushed an equal volume of mud out the top of the annulus), the downward pressure at the end of the pipe is:

$$1,500 \times 8.6 \times 0.052 = 671 \text{ psi}$$

The downward pressure at the same point from the mud in the annulus is:

$$1,500 \times 13.5 \times 0.052 = 1,053 \text{ psi}$$

Because the end of the pipe is open, the two fluids are connected, and the pressure from the mud in the annulus will push upward against the seawater in the pipe, overwhelming its downward pressure. The effect will be a differential pressure reading at the top of the pipe of 382 psi. The diagram shows the mechanism by which this occurs. The U-tube effect may have played a role in the confusing pressures seen by the *Horizon* crew during the displacement to seawater and the last minutes before the blowout.

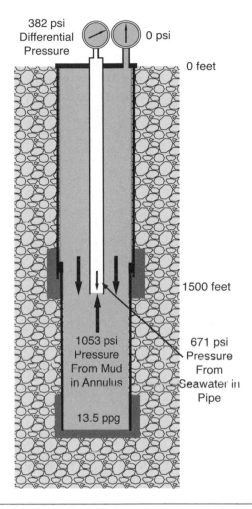

382 psi
Differential
Pressure

0 psi

0 feet

1500 feet

1053 psi
Pressure
From Mud
in Annulus

671 psi
Pressure
From
Seawater in
Pipe

13.5 ppg

Figure TN1.2. U-tube effect.

TECHNICAL NOTE 2: THE BLOWOUT PREVENTER

Cameron International Corporation manufactured the *Horizon*'s BOP, which came in two sections that were designed to separate in an emergency. The upper section contained the two annular preventers and control units, and the lower section contained five barriers called *rams*.[4] (See Figure TN2.1.) If separated, the barriers in the upper section would remain attached to the riser and could be used to prevent its load of oil-based mud from pouring

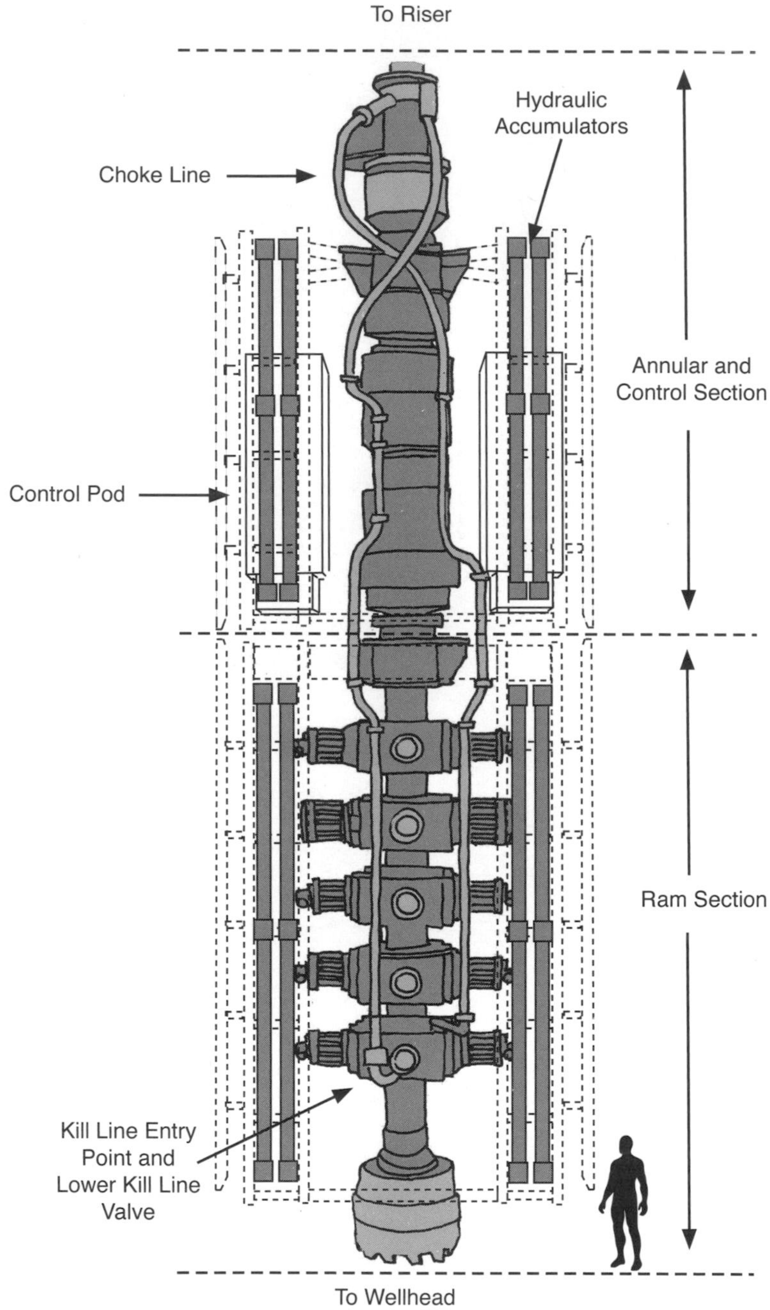

To Riser

Choke Line

Hydraulic
Accumulators

Control Pod

Annular and
Control Section

Ram Section

Kill Line Entry
Point and
Lower Kill Line
Valve

To Wellhead

Figure TN2.1. The BOP.

out into the ocean. Meanwhile the barriers in the lower section could remain on the well and keep its hydrocarbons under control.

The barriers in both sections obtained their energy from *hydraulic accumulators.* These pressure vessels operated much like the flash unit on a modern camera, charging up slowly and then releasing that energy in a pulse when needed. Each section of the BOP had its own set of accumulators.

Another set of piping ran from the rig to the bottom of the BOP, bypassing barriers called the *kill line* and the *choke line.* The kill line had one entry point below all the rams. The choke line had two entry points, one below each variable-bore ram. Each line had redundant valves at the entry point, controlled from the BOP panel in the drill shack.

A crew uses the choke and kill lines to recover from a kick after shutting in the well, usually by closing one or both of the annular barriers. Kick recovery requires circulating out the insufficiently heavy mud that allowed the kick and replacing it with higher-density mud. A simplified version of this delicate process consists of pumping the new mud down the drill pipe and allowing the old mud to return up the choke line while the closed annular barriers continue to block the riser annulus. If drill pipe is not in place, or the blind shear ram has cut it, the crew would pump the new mud down the kill line.

As we noted in Chapter 13, one or both of the lower valves on Macondo's kill line might have been closed without the crew knowing it, thereby preventing true readings of pressure and flow at the surface and misleading the crew into declaring the second negative test a success when in retrospect it clearly was not.

The Annular Preventers

The two upper barriers on the *Horizon*'s BOP—called the *upper annular preventer* or *upper annular,* and the *lower annular*—opened and closed the annulus by gripping pipe that ran down the middle. The annulars, each composed of a block of synthetic rubber, were roughly the shape of a donut. When squeezed by a conical piston, this rubber closed the pipe. The annulars could also close completely if no pipe was present (see Figure 3.3).

The annulars were used often—opened when fluids had to flow up the riser annulus and back to the rig, and closed as a precaution when fluids did not have to flow. On the *Horizon,* the crew used the lower annular for this

purpose and kept the upper in reserve. The lower annular had been modified at BP's request to a *stripping annular,* through which pipe could be pulled
when it was closed. The gain of this capability came at the cost of the ability
to hold against pressure: the lower preventer could withstand just half the
pressure the upper one could withstand.

The Rams

The lower barriers, called *rams,* closed from side to side. There were five of
them, arranged in a vertical stack, as shown in Figure TN2.2.

The topmost ram, called the *blind shear ram,* was designed to simultaneously shear and seal drill pipe. The next ram down, the *casing shear ram,*
could shear large-diameter casing but not seal it. The two rams below the
casing shear ram, called *variable bore rams,* could provide high-pressure seals
around drill pipe of varying diameters, but could not seal if no pipe was
present. The bottom ram was inverted and could seal a pipe against pressure applied from above. This was used to test the BOP.

Each ram was actuated by a hydraulic piston that, when pressure was applied, pressed the ram into the center of the bore. The pistons were in a re-
movable housing called a bonnet. The four rams intended to seal—the blind
shear, variable bores, and test ram—were equipped with devices called *ST
locks,* which automatically latched when the ram was closed. To unlock them,
the crew had to press buttons on the BOP control panel. The fifth ram, the
casing shear ram, was designed only to cut and therefore had no lock.

A Fragile Goliath

Viewed as a physical object, the *Horizon*'s BOP was reassuringly massive.
Viewed as a safety device of last resort, it was remarkably fragile.

The capability of the blind shear ram to cut through all forms of drill
pipe in all plausible situations was, as we will describe below, far from certain. Intensive activity—as could occur in a crisis—could also deplete the
hydraulic accumulators, rendering the BOP inoperative until they were
charged back up again.

The door-sized BOP panel controlled about sixty functions, with one to
three pushbuttons per function.[5] The panel was separate from the other
drilling controls and located behind the drillers' chairs, so drillers had to
get up and turn their backs to their displays to operate it.

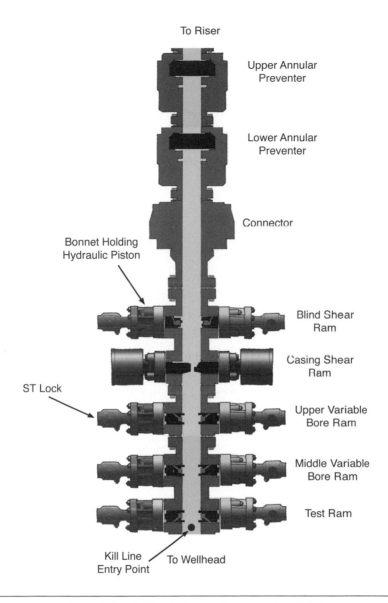

To Riser

Upper Annular
Preventer

Lower Annular
Preventer

Connector

Bonnet Holding
Hydraulic Piston

Blind Shear
Ram

Casing Shear
Ram

ST Lock

Upper Variable
Bore Ram

Middle Variable
Bore Ram

Test Ram

Kill Line
Entry Point

To Wellhead

Figure TN2.2. The closing mechanisms of the BOP.

Transocean compensated for this by requiring that two drillers be in the shack at all times, and with a de facto policy that a subsea supervisor operate the BOP controls in normal circumstances. This policy may have reduced drillers' proficiency when they confronted an emergency and no such specialist was available—as occurred on the *Horizon.* The panel did not display the status of the hydraulic accumulators, so whoever was operating the BOP had to manually estimate the pressure remaining in the accumulators based on the actions that had been commanded.

Less than three weeks before the blowout, a BP supervisor expressed concern about drillers' proficiency with the BOP controls: "As most of you are aware, we have recently had two inadvertent closings of the Blind/Shear rams and a [bottom hole assembly] inadvertently stripped into a closed annular preventer. The Blind/Shear events resulted in equipment damage. However, actuating the wrong BOPs in an actual well-control event could be catastrophic. These types of errors are totally unacceptable." Another supervisor responded that the situation was worse than the original complainant had expressed: "We have a long history of closing the wrong preventor on the wrong tool . . . All of these incidents were the result of an assumption or an act committed with no thought. There is little or no formal education on do's and don't around proper operation, testing and maintenance of well control equipment."[6]

The BP and Transocean crews were very familiar with the annular preventers, because they used them often. However, the drillers rarely used the rams. They were intended for operations such as the shearing of drill pipe during an incipient blowout—a situation that even the most senior members of the crew had never faced.[7]

The BOP Control System

Two *pods* in the upper section of the BOP controlled the assortment of annulars, rams, and accumulators. These pods, known as the *blue pod* and the *yellow pod,* contained complex systems of electronic and electromechanical components powered by batteries. A mile of electrical cables ran outside the riser from the BOP to the rig to provide control, and an equal length of hydraulic lines provided pressure to charge the accumulators or initiate some functions directly.

The basic function of the control pods was to convert the electronic signals that came down the cables into force on pistons attached to the annular preventers or the rams. The pods did this by applying electrical power to solenoids—a kind of linear actuator. The electric power produced a magnetic field that would move a plunger attached to a hydraulic valve.

Opening the solenoid-operated valve would release low-pressure hydraulic power to a second valve, forcing a "shuttle" to slide back and forth in its bore. Moving the shuttle to the "open" position would allow high-pressure hydraulic fluid to flow to the conical pistons in the annular preventers, the cylindrical ones attached to the rams, or the valves on the choke and kill lines.

Each valve attached to the choke and kill lines was equipped with a "fail-safe" spring, which moved the shuttles into a predefined position if they were in the opposite position and lost hydraulic power. This worked because these valves had no locks: constant hydraulic pressure was required to hold the shuttles in anything other than the fail-safe position. For the valves on the lower kill line, this was the "closed" position. If the crew selected "closed," both hydraulic and spring pressure would hold the shuttle in place. If the crew selected "open," the valve would stay open only as long as hydraulic pressure was applied to the shuttle. This control logic means that one or both of the lower kill line valves at Macondo might have closed as the consequence of a loss of hydraulic pressure or electrical malfunction of the first, solenoid-operated valve, and the "open loop" nature of the control system would insure that the closing took place without the knowledge of the crew.

The BOP's original design did not include the use of a fail-safe spring. In that original design, valves moved to their fail-safe positions hydraulically, and crews used a separate control panel to choose which position for a given valve was "fail-safe": open or closed. This panel was removed and the springs installed in August 2004.[8] None of the post-disaster analyses of the BOP examined the impact of this change, or the possibility that the control of the lower kill line valve had malfunctioned.

Another characteristic of the control system worked to prevent the crew from realizing that that valve was closed. The control pods operated as an "open loop": they did not have sensors that could confirm that a commanded action—such as the movement of a ram, or the opening of a lower kill line valve—had actually occurred. Instead, the crew had to rely on other indicators in the circulating system, such as changes in pressure, to tell them about the state of the BOP after they had commanded an action. In the case of the lower kill line valve at Macondo, the signal was a 6,000-psi spike in kill line pressure the first time the crew started a pump against a valve that they almost surely thought was open but in fact was closed. When the crew repeated this sequence later, the result was another 6,000-psi spike plus a disabled pump.

Although the BOP had two control pods, the control system was not fully redundant. Only one pod, selected by the crew, was connected to the hydraulic accumulators at any given time. When an electronic signal for an operation went down the cable, the electronic logic in both pods responded, but only the one with hydraulic power could initiate a physical action.

The least-certain facility of the BOP was the automatic sequencing used to shut in the well and disconnect the *Horizon*'s riser in an emergency. As noted in a technical paper written while the *Horizon* was being designed,[9] the primary sequence for this procedure was manual, to be used if the rig could no longer maintain its position over the well, lost power, or faced severe weather. If the crew did not have time to complete the sequence manually, it could activate the emergency disconnect sequence (EDS)[10] by pushing a single button on any of several control panels on the rig. This would initiate a program inside one of the control pods that would perform the sequence.

A built-in feature of the pods called the *automatic mode function* (AMF), or *deadman* system, provided a third option for disconnecting the riser from the BOP and shutting in the well. The designers of such a mechanism face the same dilemma as those who design alarm systems: they can bias their design so that it never produces a false alarm, and accept the possibility that the alarm might not sound when it should, or they can bias the design to ensure that the alarm will always sound in a true emergency and accept the possibility of false alarms.

In the case of the BOP, a false alarm would occur if the AMF initiated the shear rams when there was no emergency and the drill pipe was rotating. If that occurred, equipment could be damaged, crew members could be injured, or the drill string could drop into the well, requiring expensive retrievals or other lost-time situations. The designers of the AMF biased it in the other direction, requiring that the electrical lines between the pods be dead and all hydraulic lines show no pressure before the system could initiate the disconnect sequence. That meant that any stray electrical signal or residual hydraulic pressure would prevent the AMF from functioning even if the rig was on fire.

The final automatic function was purely mechanical: a pin in the joint between the two sections of the BOP would break if the BOP separated from the riser, initiating the blind shear ram. This was a last-ditch measure designed to seal the well while allowing the contents of the riser to escape.

How the *Horizon*'s BOP Failed

Because of the lack of survivable event-recording facilities on the *Horizon*'s BOP, we will never know the precise sequence of events that led the blind shear ram to fail to cut the drill pipe and seal the well. The point at which the ram was activated is in dispute. Various experts have asserted that it could have occurred at any time from the moment of explosion, when the AMF could have operated in response to the destruction of the cables, to the next day, when a remotely operated vehicle swam down to the BOP and cut a hydraulic line, to two days later, when another vehicle cut the mechanical pin at the juncture of the upper and lower sections. The latter scenario is the most likely, as later analysis showed that one pod had a miswired solenoid and the other had a dead battery, rendering the AMF logic in both pods inactive.

What is not in dispute—because of the physical state of the BOP and the drill pipe trapped in it when they were retrieved after the well was killed—is that the drill pipe was off-center in the blind shear ram. The design of the ram was such that this prevented the cutter on the ram from fully closing and severing the pipe. The reason the pipe was off-center is also in dispute. Some experts assert that it buckled when the heavy top drive fell from the oil-drilling derrick after the flames from the burning well had weakened the cables holding it. Other experts contend that the velocity of the gasses flowing up the riser pushed the pipe off center. Still others maintain that the pipe twisted from a pressure differential. A final theory is that it moved off center as the *Horizon* drifted—dead and on fire—off the well.

Issues Revealed after the Blowout

The *Horizon*'s BOP came under intense scrutiny after the event, almost all of it focused on the failure of functions that were supposed to seal the well after the blowout. Expert witnesses who submitted reports in support of various parties during MDL 2179 uncovered issues related to the definition, design, and maintenance of the rig's various systems. Our goal here is not to choose between conflicting expert opinions, but instead to summarize them and the conclusions of other investigators to deepen the understanding of this component.

Definitional Issues

A bedrock document used to design complex, safety-critical systems in other engineering domains such as aerospace or weapons systems is the *Concept of Operations* (ConOps), as defined in a standard systems engineering handbook: "A Concept of Operations (ConOps) document is produced early in the requirements definition process to describe what the system will do (not how it will do it) and why (rationale). It should also define any critical, top-level performance requirements or objectives (stated either qualitatively or quantitatively) and system rationale."[11]

No such document or equivalent overall specification for the *Horizon's* BOP has surfaced. As a result, any party can assert any of a wide range of assumed functions and capabilities for the BOP without fear of contradiction. As the apocryphal maxim has it, "A system without a specification can never be wrong, only surprising."

The lack of a comprehensive, reviewable document does not mean that "top-level performance requirements or objectives" for this system did not exist. Rather, it means that they were made indirectly through design choices, and no evidence exists that they underwent any form of contextual review. The result was a series of surprises, the final one being tragic.

Design Issues

The *Horizon's* BOP was initially configured around 2000 by an exploration company called Vastar Resources, which was spun off from but controlled by ARCO in 1993 and then absorbed into BP when it acquired ARCO in 2001. Vastar was to be the first company to lease the *Horizon,* a role that transferred to BP with the acquisition. BP reviewed and reaccepted the BOP configuration in September 2009.[12]

The BOP consisted of two annular preventers and five rams, as noted. Vastar decided against installing an acoustic trigger, which could have activated the rams without involving the cables, deciding instead on using EDS logic to activate the BOP's emergency sequence. This logic involved closing only the blind shear ram, rather than first cutting whatever pipe was running through the BOP with the casing shear ram and then sealing the bore of the BOP with the blind shear ram. BP implicitly concurred with these decisions in 2009 by failing to change them.

The year 2000 marked the early days of high-temperature, high-pressure wells in the Gulf of Mexico, and some decisions that were reasonable then proved inadequate ten years later. The *Horizon*'s BOP underwent three sets of modifications during that time—two of which tended to reduce its effectiveness, and one whose effect has gone unexamined by the various investigations into the blowout.

In the first modification, in November 2004, the third variable bore ram at the bottom was turned upside down so that it could resist pressure from above. This was done to improve the efficiency of BOP tests, which involved pressurizing the riser. In July 2006 the lower annular preventer was converted to a "stripper" configuration, which permitted drill pipe to be moved through it when it was closed. This change allowed faster drill pipe handling but reduced the amount of pressure the annular preventer could withstand. And in August 2004 the "fail-safe" mechanisms in all the valves were changed, a modification whose effects are currently unknown.

By the time of MDL 2179 in 2013, multiple expert witnesses for various sides in the trial defended or called into question these and other design decisions. There were three broad areas of dispute: the overall configuration of the rams, the configuration of the blind shear ram, and BP's failure to order the installation of new technology on the BOP as it became available.

Ram Configuration
The inclusion of just one blind shear ram in the set of rams eliminated redundancy and permitted the failure of just one component to derail the emergency operation of the BOP. After the Macondo blowout, the industry upgraded its recommended practices to include the use of two blind shear rams, and the successor to MMS is considering turning that recommended practice into a regulation.[13]

Expert witnesses challenged the amount of pressure the rams and annular preventers would have to cope with in a worst-case scenario. That amount is based on an estimate of the pressure in the bore of the BOP during an incipient blowout—known as the *maximum anticipated surface pressure,* or MASP. The most conservative calculation, called "gas-to-surface," is based on the assumption that the well is 100 percent full of gas when emergency action is required. One expert asserted that BP used a weaker assumption: that the well is filled 50 percent with gas and 50 percent with drilling mud. The result is a lower—and in the expert's opinion inadequate—requirement for the pressure that rams and preventers would have to withstand.[14]

Configuration of the Blind Shear Ram

The cutting elements of the blind shear ram are diagrammed in Figure TN2.3, as if the observer is looking down into the bore of the BOP from above.

Hydraulic force on the pistons in the bonnet pressed together a V-shaped shearing blade and a straight shearing blade. The blades were arranged so that the straight blade would slide under the V-shaped one, tensioning and then shearing the drill pipe. Gaskets on the blade bodies would then seal the bore when the blades were closed. However, the V-shape blade spanned only 80 percent of the bore. That means that if the drill pipe curved over to the side, as in our diagram, the blade would not apply enough cutting area to the pipe to shear through it.

Experts retained by BP put forth a variety of arguments in defense of this design, mostly centered on the idea that the crew should have actuated the ram sooner than they did, when the drill pipe was centered. Other experts criticized the EDS sequence selected by Vastar/BP rather than the ram design, asserting that if it had activated the casing shear ram before the blind shear ram, the sequence would have centered the pipe.

Failure to Upgrade

Perhaps the sharpest criticism of BP's management of the *Horizon*'s BOP centered on the company's decision to allow a circa-2000 technology package to

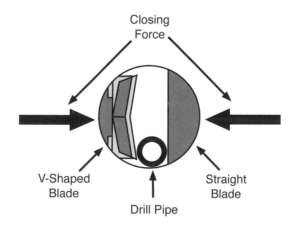

Figure TN2.3. The offset pipe effect.

remain in service for a decade, without an upgrade, while water depth, drilling depth, well pressures, and well temperatures were all increasing. One potential improvement, the acoustic trigger, was available from the beginning.

Perhaps the most obvious improvement, in retrospect, was the availability of "double-V" shearing blades instead of the "V plus straight" that was installed. The double-V configuration would have covered the entire bore, and increased the shearing force to deal with larger-diameter pipe or tool joints, the reinforced sections where lengths of pipe screwed together. BP also declined to install "tandem boosters," a double-piston form of bonnet that would also have increased shearing force, as would have raising pressure in the hydraulic accumulator banks to 5,000 psi, which BP also failed to do.

Finally, Cameron had developed a new control system with features such as the ability to recharge the batteries in the pods. An upgrade to this system was planned for the *Horizon*'s stay in dry dock in 2011, but the blowout came first.

Maintenance Issues
Transocean applied a variation of its "condition-based maintenance" philosophy—which its chief mechanic derided as "run it until it breaks"[15]—to the upkeep of the BOP. Transocean's subsea team called this approach "condition-based monitoring." Instead of waiting for something to break, the team would wait until the BOP was pulled off the bottom when a rig was about to move off a well, partially disassemble it and inspect various parts, mainly gaskets and seals for wear, and replace the ones that looked bad. This approach contrasts with periodic maintenance, in which technicians tear down and completely rebuild the BOP's engine at specified intervals. As a result, some unknown number of BOP components had not been tested or replaced for ten years, and the incremental attention to the BOP inhibited the installation of upgraded features.

By contrast, Cameron itself, the American Petroleum Institute's Recommended Practice 53 (RP 53), and MMS regulation 30 C.F.R § 250.446 (which references RP 53) all called for tearing down and inspecting the BOP every three to five years. The draft JIT reports interpreted these documents in a way that implied Transocean had violated them—an implication that Transocean vigorously rejected.

Transocean offered both substantive and legal defenses of its policy. The substantive defense was that condition-based monitoring is actually superior to periodic maintenance because it avoids "infant mortality"—the increased chance that newly-installed parts will fail.[16] The legal argument was that RP 53 was worded as a recommendation and not a requirement, so the MMS regulation was advisory only.[17] The JIT was not impressed, and the citation remained in the Department of the Interior report, although there is no evidence that fines were imposed or other enforcement actions took place for this.

Whether condition-based monitoring is or is not superior to periodic maintenance, it failed to detect that the battery in the blue pod of the *Horizon*'s BOP was dead, and that a critical solenoid in the yellow pod was miswired. Those problems prevented the automatic disconnect logic from activating the blind shear ram during the Macondo disaster. What is not known, considering the number of issues associated with that ram, is whether the ram would have sealed the well even if the EDS had operated as designed.[18]

TECHNICAL NOTE 3: DRILLING MARGIN

In Chapter 5 we introduced the concept of drilling margin—the amount of reserve push (hydrostatic force) drillers can apply to counter unexpected increases in shove (pore pressure) without breaking down the formation and losing mud into it, thereby reducing push to a dangerous degree. In this note we explain how drillers use the formulas of primary well control to calculate drilling margin, and how they cope with a loss of it by running casing and then testing to see whether they have a safe margin (see Figure TN3.1). To do this, we will expand on the example we used to present the formulas in Technical Note 1.

As before, we have a porous zone at 1,000 feet with a pore pressure of 12.0 ppg, and a well hole held in place by 12-ppg mud. Let us further assume that this porous zone is in a fragile formation that has been previously tested (by methods we will describe below), and is known to break down at a fracture pressure of 12.5 ppg. At this point the drill crew is operating at the minimum drilling margin without special permission from MMS.

If it turns out (in our unrealistic, hypothetical well) that the testing was inaccurate, and the pore pressure, as determined by measurement-while-drilling equipment, is 12.1 or 12.2 ppg, the drill crew would be required, by regulations in place in 2010, to shut in the well and request permission

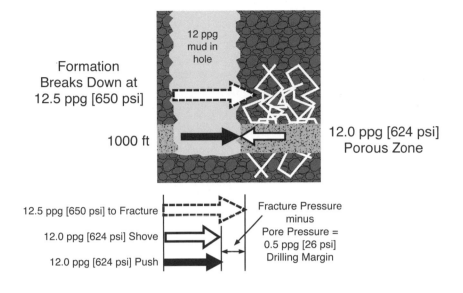

Figure TN3.1. Drilling margin.

from MMS to drill at or near the absolute minimum legal margin of 0.3 ppg. The crew could then "weight up" by circulating out 12-ppg mud and replacing it with 12.1- or 12.2-ppg mud.

If the crew encounters a later porous zone that narrows the drilling margin further, they will have to run casing and then test the formation to determine the new drilling margin. To explain this, we will extend our hypothetical well down to 2,000 feet (see Figure TN3.2).

Here the crew has drilled down to 2,000 feet and encountered another porous zone of 12.3-ppg pore pressure, requiring them to "weight up" to 12.3-ppg mud in the hole. This, in turn, has reduced the drilling margin up at 1,000 feet to 0.2 ppg, requiring the crew to run casing. The crew uses a float collar and wiper plugs to run casing into the hole and cement it at the bottom, as we described in Chapter 7. At the end of the process our hypothetical well would look like the diagram shown in Figure TN3.3.

The possibility of lost returns at 1,000 feet or an influx of hydrocarbons at 2,000 feet has been forestalled by the combination of casing and cement, and the crew can weight up to a heavier mud based on geologists' predictions of what they will encounter when drilling the next interval. However, before they press ahead they must test the formation below 2,000 feet to verify that

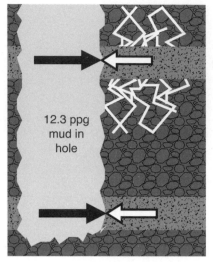

1000 Foot Depth
12.3 ppg [640 psi]
Push

12.3 ppg
mud in
hole

2000 Foot Depth
12.3 ppg [1279 psi]
Push

12.0 ppg [624 psi]
Porous Zone in
12.5 ppg [650 psi]
Fragile Formation;
Drilling Margin =
0.2 ppg [10 psi]

12.3 ppg [1279 psi]
Porous Zone

Figure TN3.2. Running out of drilling margin.

the formation can support the weight of mud they plan to use at that depth. They begin the test by drilling through the bottom cement and down a short distance, as shown in Figure TN3.4.

If the crew tested whether the next interval of hole could withstand the planned increase in mud weight simply by circulating heavier mud into the hole, but were wrong about the strength of the formation, they would immediately face lost returns. Instead, they simulate the effect of heavier mud by applying external pressure to the mud already in the hole. Because formation strength usually increases with depth, the top of the new interval should be (but not always is) the weakest point.

Crew members then perform a *formation integrity test* (FIT) or a *leak off test* (LOT). Both involve slowly adding pressure to the mud column using a low-volume, high-pressure pump, such as that used to pump cement into the well. The two tests differ in structure. In the case of the FIT, the crew uses the pump to apply a selected pressure to the mud. If a fracture does not occur, the crew can safely use that pressure in that section of the well. If a fracture does occur, that pressure could result in lost returns and eventually a kick. In an LOT, the crew applies pressure until the formation fractures. The pressure at that point is unsafe—it will cause lost returns. In both tests the crew converts the pressure to an equivalent mud weight.

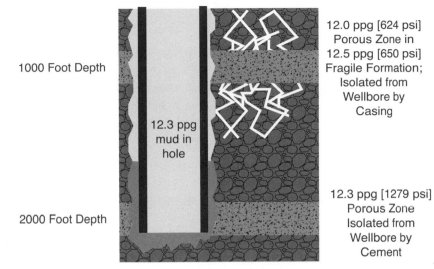

1000 Foot Depth

12.3 ppg mud in hole

2000 Foot Depth

12.0 ppg [624 psi] Porous Zone in 12.5 ppg [650 psi] Fragile Formation; Isolated from Wellbore by Casing

12.3 ppg [1279 psi] Porous Zone Isolated from Wellbore by Cement

Figure TN3.3. Running casing.

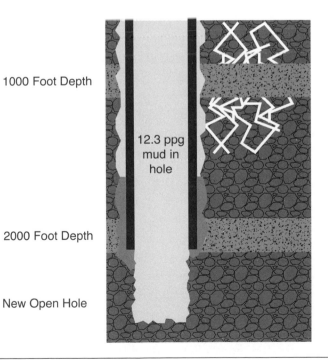

1000 Foot Depth

12.3 ppg mud in hole

2000 Foot Depth

New Open Hole

Figure TN3.4. Drilling ahead.

Crews can use either test to validate a decision to circulate out lighter-weight mud and circulate in heavier mud. Both tests require geologists who interpret pressure curves while pressure is rising and then as it is held steady for some period of time.

A plaintiff's expert witness asserted that the *Horizon* crew did not use pressure sensors in the drilling hole, relying instead on less-accurate pressure measurements at the surface. He also asserted that the crew misinterpreted the results of three of five sets of tests at Macondo, misrepresented all the results to federal regulators, and drilled ahead without the required drilling margins.[19] BP's experts challenged these conclusions.[20] The exchange exposed the degree of interpretation required by both the regulations and the tests.

TECHNICAL NOTE 4: CEMENTING MECHANISMS

In the interests of simplicity in the main text, we treated two key mechanisms on drilling equipment in abstract form: the float collar and the mechanism for releasing wiper plugs, which protected the "slug" of cement as it moved down through the mud-filled casing on its way to its intended location, such as sealing the production casing at the bottom of the well. Float collars, and the spring-loaded nonreturn valves they contain, are designed not to resist pressures produced by hydrocarbon flow but instead to ensure the smooth and proper passage of cement slurries.

The operation of the float collar and its valves is significant in the case of Macondo because the device may not have performed as designed. Both it and the mechanism for releasing the plugs are also of interest as examples of how the technology of well construction uses the flow and pressure of fluids to both sense and control the state of elements thousands of feet underground.

The Float Collar

As noted in Chapter 12, the float collar, screwed into the casing, acts as a convertible valve with two states. First it permits bidirectional flow, and then, after it is commanded to convert, it restricts flow to downward only. The device performs these functions by purely mechanical means.

The left-hand diagram in Figure TN4.1 shows the general arrangement of the Weatherford float collar used at Macondo. It has two spring-loaded flapper valves. When closed, these block flow in the direction opposite to

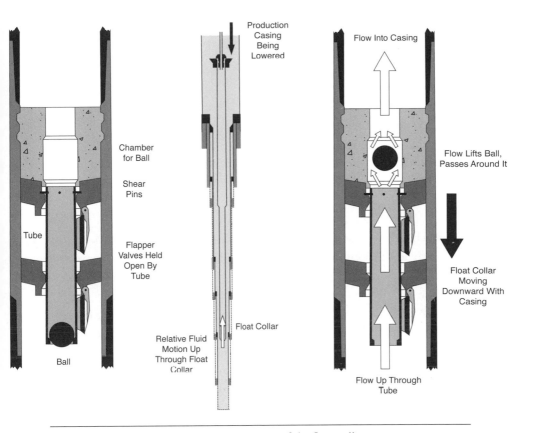

Figure TN4.1. Operation of the float collar.

the flow of the cement slurry, which moves down into the shoe track and up into the annulus and around the casing. If the valves are closed while the production casing is being lowered, as shown in the middle diagram, they prevent mud from entering the bottom of that casing, and as a result produce surge pressures against the formation—which in Macondo was very fragile.

To permit upward movement of mud into the casing, the float collar in its unconverted state has its valves propped open by a hollow tube. The tube is held in place by deliberately weak brass shear pins. A free-floating ball is in the tube. The ball is prevented from falling out the bottom of the tube by a narrow neck, and can float upward into a wider chamber but no farther.

When the casing and the float collar are moving downward while being lowered in place, the relative motion of the assembly causes the ball to float upward into the wide chamber, which in turn permits mud to pass around it and into the body of the production casing.

Conversion

After the production casing has landed and is resting on the wellhead, downward movement stops and the float collar is ready for conversion. The crew commands this change of state by setting the pumps to produce a specific pressure and flow rate, which affects the float collar as shown in Figure TN4.2.

The crew selects flow and pressure sufficient to push the ball into the end of the tube, and then to place enough stress on the shear pins to cause them to break away. The ball and tube are then pushed clear of the valves—as shown with the upper valve in the left-hand diagram—and then they drop harmlessly to the bottom of the shoe track. The float collar is then converted, as shown in the right-hand diagram, to a state in which mud and later cement can flow downward but nothing can flow upward.

In trying to convert the float collar at Macondo, the crew had to increase pressure nine times before the circulating system signaled that downward flow was achieved. Many experts think the float collar was jammed with cuttings and other debris stirred up when the casing was run, and that it never converted. Instead, this theory holds, the ball blew out the end of the tube, leaving the device in an unknown state with unknown effects on how the cement slurry actually flowed into the shoe track.[21]

Wiper Plugs

The mechanisms that manage the wiper plugs solve the problem of how to place a wiper plug ahead of and behind cement slurry when the pipe is too small to accommodate the plugs. The solution is to use a running tool, which is triggered by actuators called *darts*.

Bottom Plug

The plugs and their associated running tool are placed in the production casing before it is lowered in place. The three elements form a "stack," as shown in Figure TN4.3.

Just before the liquid cement slurry is pumped into the drill pipe, a dart made of flexible material is inserted through a special fixture. This bottom dart is pushed down the drill pipe ahead of the cement slurry, as shown. Then

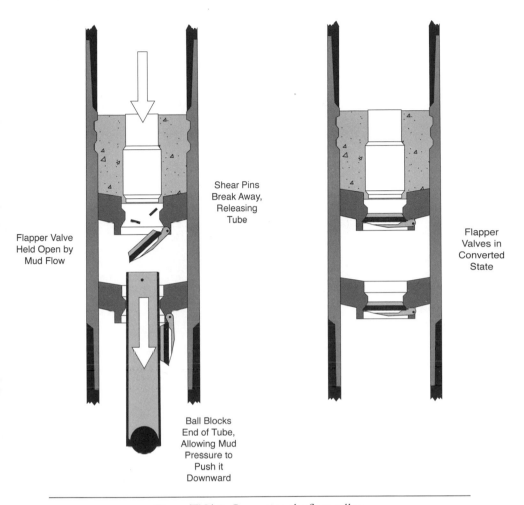

Flapper Valve
Held Open by
Mud Flow

Shear Pins
Break Away,
Releasing
Tube

Flapper
Valves in
Converted
State

Ball Blocks
End of Tube,
Allowing Mud
Pressure to
Push it
Downward

Figure TN4.2. Converting the float collar.

the dart is pushed through the running tool and the top wiper plug, engaging the bottom wiper plug, as shown in Figure TN4.4.

When the dart engages the bottom plug, it triggers a release mechanism that separates it from the top plug. The bottom plug and its dart are then pushed down by the cement slurry being pumped in behind them, as shown in the left-hand diagram.[22] The mud preceding the plug forces open the spring-loaded valves in the float collar, which allow that mud to go down the shoe track, back up the annulus between the production casing and the formation, and up the riser.

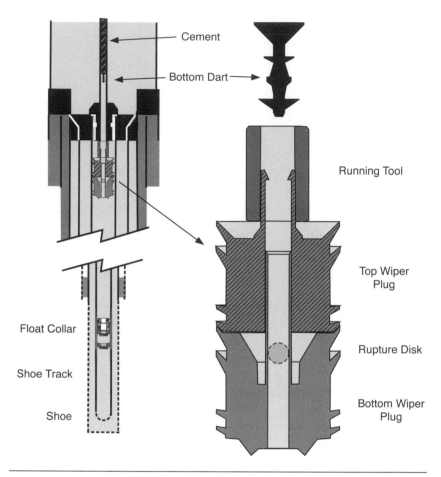

Figure TN4.3. Inserting the bottom dart.

Top Plug

After the last of the cement has been pumped, the crew inserts a second dart into the pipe, and pumps drilling mud into the pipe to push the dart down. When this dart engages the top plug, as shown in Figure TN4.5, it releases the plug from the running tool, and the combined bottom plug–cement slurry–top plug is pushed down the casing.

When the bottom plug hits the float collar, the rupture disc in the plug bursts. That allows the cement slurry to flow down through the plug, into the float collar, through the valves in the float collar, and down into the shoe track. The slurry eventually flows back up into the annulus between the production casing and the formation, as shown in Chapter 7. When the top

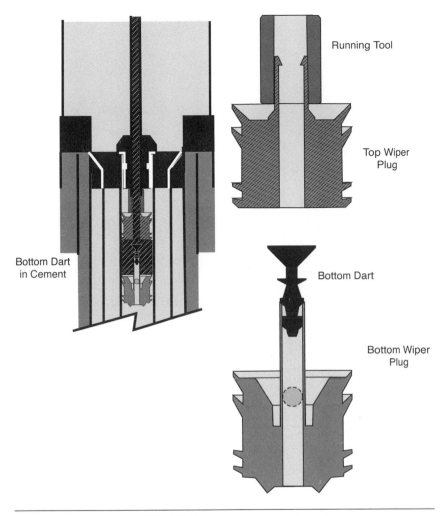

Running Tool

Top Wiper
Plug

Bottom Dart
in Cement

Bottom Dart

Bottom Wiper
Plug

Figure TN4.4. Releasing the bottom wiper plug.

plug arrives and rests on the bottom one, all the cement slurry is in the shoe track, and the annulus and the cementing operation are complete.

Each transition—such as the release of a plug, and the contact between the bottom plug and the float collar—sends a distinctive signal up through the circulating system in the form of changes in flow or pressure, which allows the crew to assess the progress of the cementing activity. After the crew cemented the bottom of the production casing at Macondo, these signals were normal except for rate of flow. This discrepancy could have led a cautious team to call for performing the cement bond log test, especially as the

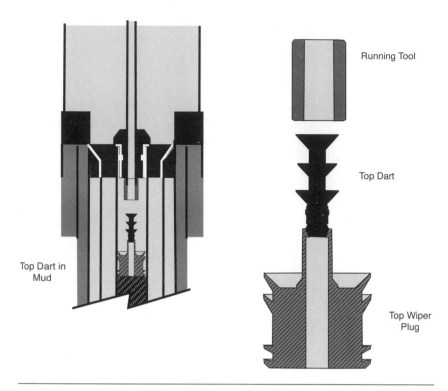

Figure TN4.5. Releasing the top wiper plug.

personnel and equipment to do the test were standing by on the rig. Instead, perhaps to avoid spending eight to ten hours on that test, the Macondo team declared the cementing successful.

TECHNICAL NOTE 5: LOST RETURNS DURING DISPLACEMENT

Despite extensive efforts by forensic analysts, no one has put forth a generally accepted explanation of what occurred at the bottom of the well to produce the signals recorded on the Sperry Sun charts. Even one of the senior WSLs, who had intimate knowledge of the *Horizon* and decades of experience, was baffled:

> [The BP investigators] asked me to come in for a couple days and try to figure out what happened to the rig in the last 30 minutes. There was some anomalies on the [Sperry Sun] charts that were unexplained.

And after three days of looking at and trying to figure it out, we left and it was still anomalies on the chart that weren't explained. We couldn't figure out what happened.[23]

At least three attempts have been made to apply computer simulations to the problem. The first was commissioned by Transocean,[24] the second by BP,[25] and the third by the Chemical Safety Board.[26] However, all these efforts assumed unrealistically low efficiencies for the main pumps, or produced results that were inconsistent with eyewitness evidence.

A major confounding factor for both the crew and later forensic analysis has been the likelihood that the crew confronted a well-control situation involving lost returns followed by a kick rather than a simple kick. The crew may not have considered lost returns when analyzing signals from the circulating system, because the positive pressure test had indicated that the production casing was sound enough to prevent the flow of mud into the formation. An independent drilling consultant first proposed the possibility that the two displacements produced lost returns in an analysis he posted to the Internet.[27] The Chief Counsel's Report downplayed this possibility,[28] but consultants to the Chemical Safety Board accepted it.[29] Here we describe the evidence that supports this hypothesis, and its implications.

The Positive Pressure Test

The crew performed the positive pressure test between 10 a.m. and noon on May 20, using the cement pump lined up on the kill line, as shown in Figure TN5.1.

The crew pulled the drill pipe above the BOP and closed the blind shear ram to seal off the well. The crew then applied pressure through the kill line, which enters the BOP at the bottom, below the annular preventer and all the lower rams. This use of the kill line raises the possibility, as noted in Chapter 13 concerning other actions, that the valve at the bottom of that line may have been closed—either inadvertently or through malfunction of a control—without the crew's knowledge. If that was the case, pressure would have been isolated from the well, and the test would have been invalid.

Possible Changes at the Bottom

At the bottom of the well, cement may have broken down or shifted, or the change in temperature between the hot drilling mud in place and the cold

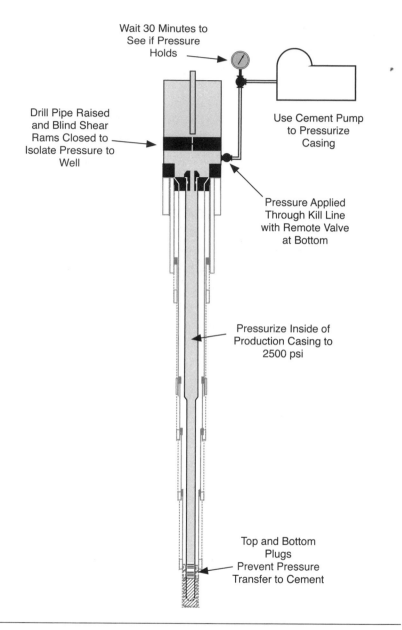

Wait 30 Minutes to
See if Pressure
Holds

Use Cement Pump
to Pressurize
Casing

Drill Pipe Raised
and Blind Shear
Rams Closed to
Isolate Pressure to
Well

Pressure Applied
Through Kill Line
with Remote Valve
at Bottom

Pressurize Inside of
Production Casing to
2500 psi

Top and Bottom
Plugs
Prevent Pressure
Transfer to Cement

Figure TN5.1. The positive pressure test.

seawater pumped in during the first displacement could have put over-whelming stress on the two and a half miles of production casing. As Chapter 7 showed, the lower portion of this casing was composed of material whose condition was unknown.[30] The casing may have buckled when the crew was running it in, as a Halliburton expert later asserted,[31] although any breach this caused would have had to pass the positive pressure test.

These possibilities combine with others that are unrelated to the positive pressure test. The float collar may or may not have converted, and cement slurry may or may not have been distributed properly down the shoe track and up into the annulus between the production casing and the formation. The nitrogen-foamed cement may or may not have attached itself to the formation, and the slurry may or may not have solidified in either the annulus or the shoe track. The higher pore pressure of the misinterpreted M57B strata may have invalidated multiple assumptions underlying the design of the cement job. The basic formula of systems analysis suggests that the bottom of the well could have been in any of a large number of states—the majority of which would have allowed hydrocarbons to flow out of and then into the production casing.

As a result, we should be resigned to the possibility that we may never have a narrative that cogently incorporates all the evidence on what was going on inside Macondo during its last hours.

Losses during the First Displacement

We can estimate lost returns during the first displacement because we have information that was not directly available to the crew: the amount of mud transferred to the *Bankston* between 1:28 p.m. and 5:17 p.m. on May 20. In testimony before the Marine Board of Investigation, the captain of the *Bankston*—an experienced mariner who showed a high degree of professionalism that night, and who could later refer to his logbook—reported that his crew transferred "approximately 3,100" barrels of mud during that period.[32] Because the material affected the trim of his vessel and also had to be accounted for as toxic waste, we can assume that this estimate is valid.

We can then calculate what the change in pit volume *should* have been, given "full returns"—that is, no loss of mud into the formation. The diagram in Figure TN5.2 summarizes the simultaneous flows into and out of the pits.

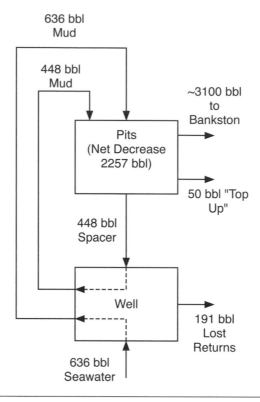

Figure TN5.2. Lost returns during mud transfer to the *Bankston*.

The diagram is organized by the type of fluid being moved. The crew had pumped 450 barrels of spacer into the well, which under our "full returns" assumption would have displaced 450 barrels of mud out of the drilling pipe and into the pits. The crew also pumped in 636 barrels of seawater at different times, to displace mud in the auxiliary lines of the BOP, and to push the spacer above the annular preventer. This should have displaced an equal amount of mud into the pits, increasing their volume by 636 barrels, for a total increase of 1,084 barrels.

Examination of the Sperry charts shows that the measured volume in the pits *dropped* by 2,257 barrels from 1:30 p.m. to 3:20 p.m., a decrease that includes the 448 barrels of spacer. (We chose that time period because the pit volumes are stabilized at the beginning and end of it.) During that period, 1,084 barrels were flowing into the pits. In all, that means we must account for 3,341 barrels of mud.

Sperry data show that the crew pumped 50 barrels out of the pits and into the riser to "top it up" when they noticed the drop in fluid in the riser. About 3,100 barrels went to the *Bankston*. This leaves a discrepancy of 191 barrels, which had nowhere to go but into the formation.

The exact amount of lost returns is uncertain because the captain of the *Bankston* rounded his number, the hose connecting the *Bankston* to the *Horizon* may have contained residual mud, and converting the curves of the Sperry charts into numbers is imprecise. However, the notion that about 100 to 200 barrels of mud was lost during the first displacement is supported by the measurement the Sperry flow sensor registered on the charts for the period.

The sensor reported a higher flow than the more reliable changes in pit volume recorded during the second displacement, which we take as evidence of a higher amount of loss.[33] Hydrostatic calculations also show that larger losses can roughly account for pressure anomalies during the negative tests. Whatever the precise number, we believe the totality of the evidence supports a conclusion that lost returns during the first displacement were enough to affect the signals the crew would have seen when they attempted the negative tests.

Losses during the Second Displacement

The situation during the second displacement—from just after 8:00 p.m. until 9:08 p.m., when the crew diverted returns from the riser overboard—is much more straightforward. We can compare the increase in pit volumes during that period with the volume of seawater pumped in, as shown in Figure TN5.3.

This comparison is possible because the mud engineer had ordered the crew to stop the transfer of mud to the *Bankston* some five hours earlier. The comparison shows that about 120 barrels was lost to the formation. Even more significant is the way the losses were distributed over time (see Figure TN5.4). This chart duplicates—and, because it was independently produced, validates—the observations of a Halliburton expert witness, the analysis of InTuition Energy, and the consultant engaged by the Department of the Interior.[34]

The vertical distance between the two curves represents the total amount of mud lost to the formation. This total remains the same up to 8:40 p.m., at which time the amount lost starts to increase.

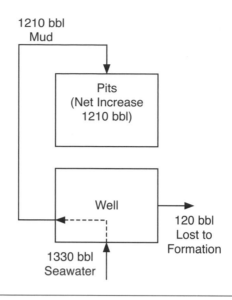

Figure TN5.3. Lost returns during displacement.

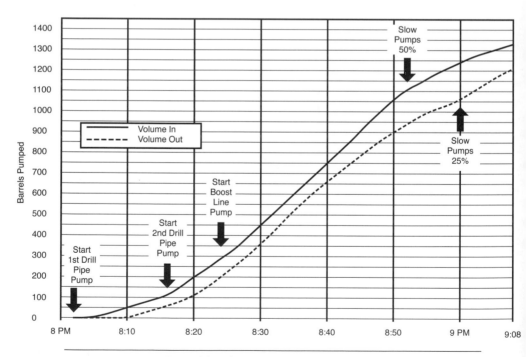

Figure TN5.4. Lost returns turn into a kick.

Just before 9 p.m. the total amount lost rose to about 180 barrels. At this point the crew had slowed the pumps, which reduced the extra push produced by the velocity of the seawater going into the well. The gap between the two curves now starts to narrow, signifying a decrease in the total amount lost to the formation. This phenomenon can only be explained by hydrocarbons entering the well and causing more fluid to be returned than is being pumped in: the classic indication of a kick. Forty minutes later Macondo blew out.

TECHNICAL NOTE 6: THE *HORIZON'S* CIRCULATING SYSTEM

In Chapter 2 we presented the circulating system in the simplified form shown in Figure TN6.1.

A complete understanding of how the characteristics of this system influenced and constrained the actions of the crew requires a more detailed exploration of both its logic and its physical configuration. Various investigations have failed to uncover a specification or detailed diagrams for its physical layout, and Transocean did not back up essential documentation on shore, so much detailed information on it went down with the *Horizon*. What we outline is, in effect, the outcome of reverse engineering based on trial testimony and knowledge of similar systems, and like all such exercises it may contain errors.

The expanded logic of the system as used on the last day is shown in Figure TN6.2. This diagram shows the crew in the drill shack backed up by the mud logger, and the cement unit connected to the drill pipe through a chiksan line, and enumerates the relevant sensor signals and commands. The sensors provided information on the volume of fluid pumped into the well and returning to the storage pits, the rate of flow from the riser into the pits or overboard, and pressure on various lines. The crew also used two classes of commands to control the well: those used to start and stop the main pumps, and those used to open and close the annular preventer and the valves on the kill line attached to the BOP. The crew also sometimes used the cement unit to control the well.

At a physical level, the *Horizon's* circulating system was a complex, ad hoc conglomeration of elements from multiple vendors, spread over four locations on the rig, and subjected to an unknown number of fixes and modifications

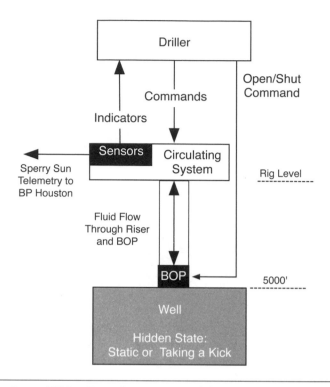

Figure TN6.1. General arrangement of the circulating system.

over its lifetime. Control was distributed among multiple decks, as shown in Figure TN6.3.

Three of the four locations—the drill shack, mud logger's office, and cement unit—had their own set of displays, some unique and some redundant with those at other locations. Only crews in the drill shack and the cement unit could use pumping to change the state of fluids in the well. The shaker room contained the sensors that measured the flow returning up the riser. The pits and main pumps were on two decks below the shaker room. The pits were equipped with volume sensors, and the pumps with counters that relayed how many strokes had been pumped during a given interval.

No facility displayed how the various elements of the circulating system were lined up. If a crew member needed that information, he or she had to deduce the lineup from sensor readings, go to the drill floor or shaker room, or ask someone who knew.

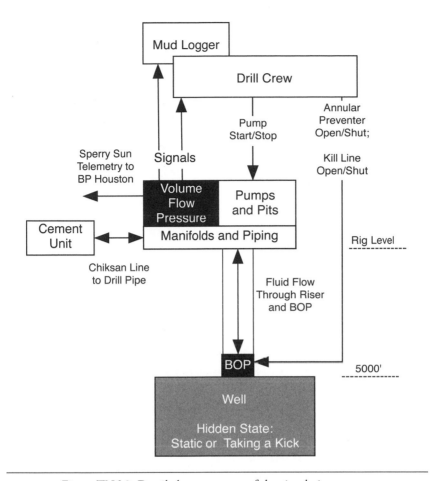

Figure TN6.2. Detailed arrangement of the circulating system.

Three different corporations staffed the four locations: Transocean, the drill shack and the shaker room; Sperry, the mud logger's office; and Halliburton, the cement unit. All were coordinated by the joint authority of BP's WSL and Transocean's OIM.

The base circulating system was installed when the *Horizon* was built. The evidence does not tell us who, if anyone, integrated the system's various elements, although the most likely candidate was HiTec ASA when it was an independent Norwegian company, before NOV acquired it. The system was then modified early in the *Horizon*'s career when Sperry Drilling

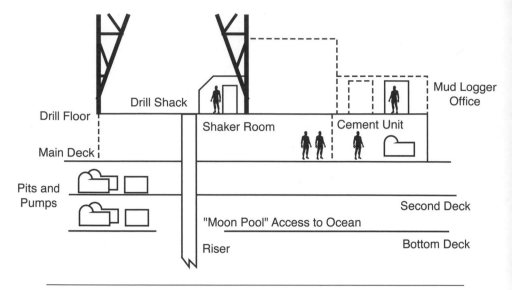

Figure TN6.3. Physical layout of the circulating system.

Services—then an independent company and now a subsidiary of Halliburton—contracted to provide mud-logging services on the rig.

The Sensor Set

Sensors on the *Horizon* measured the three basic parameters of well control: pit volume, return flow, and pressures.

Pit Volume

As we noted in Chapter 2, the most basic and reliable way to determine the state of the well during dangerous tasks is to compare the amount of fluid, such as spacer or seawater, pumped into the well with the amount of mud returning up the riser and into the pits. If more fluid is coming out than is going in, the well is taking a kick. If less fluid is coming out than is going in, the well is losing fluid into the formation.

The tremendous pressures that pumps use to force fluid into a well preclude any direct measurement of the amount flowing in. Instead, on the *Horizon,* devices attached to the pumps counted the number of pump strokes, and the Sperry and HiTec equipment converted this number into flow volumes.

To determine the amount of fluid in the pits, the driller or mud logger would observe readings from sensors that used acoustic echoes to sense the depth of the fluid. However, the motion of rigs such as the *Horizon*—known not entirely affectionately by drill crews as "floaters"—can cause readings on the pit volume sensors to fluctuate by as much as thirty barrels: two to six times the amount regarded as a kick. Such motion can stem from sea state, adjustments to trim by the marine crew, or the use of two large cranes on either side of the deck to move heavy objects around or to and from a support vessel. Crews can obtain more reliable readings on small volumes by using a small, vertical tank called a *trip tank.*

Return Flow
Two flow sensors, one supplied by HiTec and a second added later by Sperry, indicated the rate at which fluid was coming out the riser during the displacements and negative tests. The HiTec sensor used a spring-loaded paddle to sense variations in flow. The Sperry sensor used acoustic signals to determine the depth of fluid in the flow line. Both sensors were subject to fluctuation caused by rig motion.

Both sensors had limitations. The maintenance supervisor on the *Deepwater Nautilus* asserted after the blowout that the HiTec paddle-style flow sensor had "fundamental design flaws."[35] The mud logger on the *Horizon* testified that this sensor sometimes indicated that the flow of mud out of the riser was 30 percent of the flow going in when the crew knew that it was 100 percent. And because the Sperry sensor detected depth of fluid rather than motion, the sensor would show that flow was occurring even when stationary fluid had pooled in the bottom of the flow line.

Crew members could observe flow directly at points where it emerged from a pipe and dropped into one of two bins that led to the pits, one of which was monitored by a closed-circuit television camera. However, the television system was unreliable because the lens of the camera required frequent cleaning to remove fluid spatters,[36] and because the camera monitored only one of two possible paths for the flow.

Pressures
Pressure readings play a role in well control because a drop in pressure could signal that hydrocarbon gas has entered the well from the formation and a kick is in progress. On the *Horizon,* there were pressure sensors connected

to the drill pipe and on the kill line attached to the BOP. The unusual—and to this day undetermined—situation that produced the Macondo kick did not present the crew with readings indicating a drop in pressure, but other signals on the pressure gauges indicated that the well was probably not static. The drill crew did not shut in the well in response to those signals.[37]

The crew may not have regarded those pressure readings as reliable enough to require action. The junior engineer reported in his internal BP interview that when they were cementing the production casing, the crew had dismissed a pressure anomaly as stemming from a faulty sensor. Indeed, no evidence suggests that anyone had calibrated the pressure sensors since the rig was built. On the other hand, the values reported by the independent pressure sensor in the cement unit exactly tracked those reported by the sensor on the drill pipe, according to Sperry data. There is no indication that the crew noticed this correlation.

If they were concerned about the accuracy of the readings on the electronic displays, the crew could have checked mechanical pressure gauges on the drill floor. That would have entailed dispatching a floor hand to read them, who would then have radioed the values to the drillers. No evidence suggests that this occurred.

Commands

The crew issued the primary commands to the circulating system—to start, stop, and adjust the pumping rate of the four main pumps—from the drillers' chairs. These pumps could move larger volumes of fluid at lower pressures than the pumps in the cement unit. On the last day, two of the main pumps were lined up on the drill pipe, a third on the kill line, and the fourth on the boost line, an auxiliary line that entered the bottom of the riser and was used to provide extra upward force to the fluid being displaced out of the well.

The crew issued other commands from the BOP control panel. Crew members used this panel to perform some sixty functions, including opening and closing the two annular preventers, selecting the amount of pressure the preventers would apply to the drill pipe, and opening and closing the two lower valves on the kill line. Each valve had a fail-safe mechanism that would close the valve mechanically upon loss of pressure or signal.

On the *Horizon,* control of the valves was "open loop": no sensor on the valve signaled its actual position. When a crew member pressed the button on the panel for a given valve, the light would illuminate even if the valve failed to move, or if the fail-safe mechanism moved it to a default position other than the one selected. Thus, the "open" button could be lit even when a valve was closed.

Evidence suggests that such a malfunction may have occurred at least once and possibly three times during the *Horizon*'s final hours. Two of the malfunctions might have led the crew to start a pump against a closed valve, causing a pressure spike. The second time it likely happened, the spike put the pump out of commission. The third possible time—between the other two—could have led the crew to misinterpret the critical second negative test by attributing a lack of flow to a sound cement barrier at the bottom of the well. In reality, unbeknownst to the crew, a closed lower kill line valve may have caused the lack of flow.

As we noted in Chapter 11, the BOP control panel was behind the drillers' chairs, and the drillers could not observe the panel while looking at their displays. The BOP panel was also not visible to the mud logger. Although it was Transocean's practice to have a subsea specialist operate the panel during normal activities, for some unexplained reason the subsea specialist visited the drill shack only intermittently during the *Horizon*'s final hours.

HiTec Issues

NOV provided the consoles that showed the drillers critical parameters such as pressures, flows, and pit volumes. The consoles were officially named Cyberbase, but workers in the field called them HiTec, Hitech, or similar, because HiTec ASA had developed the system in the early 1990s. NOV negotiated a marketing agreement with HiTec ASA in the late 1990s, and purchased the company sometime later.

The version of HiTec on the *Horizon* dated to 2006.[38] It ran on Windows NT—an operating system that Microsoft had launched in 1996 and had stopped supporting in 2005. At some point before 2009, the *Horizon*'s HiTec systems began to stop at unanticipated times, a phenomenon the crews variously called a "lockup," a "crash," or "blue screen of death." At least two crashes occurred during well-control incidents involving lost returns—one on March 14 and the other on April 3. In the first such incident, the console at the A-Chair crashed twice, and in the second, consoles at both the A- and B-Chairs crashed.[39] The chief electronics technician recalled that a console had crashed on a previous well and interfered with the handling of a kick.[40]

The hard drives in the chair computers were replaced sometime in March[41]—a step that the technician believed was insufficient to fix the problem:

> They're—BP wouldn't allow the down time to change out the computers that needed to be—desperately needed to be changed. We were given opportunities to change out hard drives, you know, internal, which is a 10-minute process, versus changing out entire computer systems, which is what was needed. The computers that we were using were—were outdated from the day they were installed. The software that was in them was a very outdated Windows NT, and it was very unstable.
>
> And everyone knew this. It would frequently cause the Chairs to lock up or give erroneous data. And—but we simply weren't ever given the time to shut down long enough to change those computers out, get them back in sync, and get us back to working.[42]

Some drillers believed that if all the HiTec chairs crashed, they could continue to maintain well control by using a Sperry Sun display next to the A-Chair.[43] There is no evidence that the crew had tested this belief either through drills or operationally. It was definitely the practice for the drillers to move to the next chair when the one they were using crashed, and to return when the chair came back up.[44] This did not always work: "The—the kick they took in March, the Chair went down, the tag replicator did not function properly. When the Chair was brought back up, the data that the Driller was looking at was erroneous. The Assistant Driller sitting over on the B-Chair or someone else said, 'Hey, you know, look, something is going on.' They discovered there we had, in fact, taken a kick."[45]

The uncertain status of the HiTec equipment raises questions, particularly about what the crew members looking at the A-Chair display were seeing when the first mate heard someone say, a half hour before the blowout, "We may need to circulate."[46] The lack of evidence that the crew had performed any end-to-end calibration or testing after the equipment was installed or even later—which would have entailed producing sensor readings and verifying the displays against them—compounds this uncertainty. An audit of the *Horizon*'s maintenance status by BP's internal Marine Authority in September 2009 reported: "According to maintenance history calibration of critical drilling instrumentation remains an area where improvement is required. Despite previous recommendations it could not be demonstrated that all critical digital and analogue drilling instrumentation is being calibrated."[47]

The audit went on to criticize management and change-control procedures on the rig: "A robust software management system in line with BP expectations, and Transocean Operations and Maintenance Advisory Notice, with the exception of DP software, could not be demonstrated. This was reported during the last audit. Interrogation of the system highlighted omissions from both the software register and hard copy back up files for critical software. Poor system management and audit appeared to have been exacerbated due to recent changes in Electrical Supervisor."[48]

After the Macondo blowout, Town employees testified that they were unfamiliar with problems with the HiTec system, or that the problems were not severe, or that replacement of the hard drives fixed them.[49] This view contrasts markedly with that of the rig's chief electronics technician, who testified that the chair software was unstable up to the end.[50] According to a report on the last inspection of the system, performed by an outside organization eight days before the blowout: "Interviews were held with the operating and maintenance crew. Both parties were satisfied with the HiTech cyberbase [*sic*]. The software on any of the three chairs was stable and had not shown (excessive) crashes."[51]

The word "excessive" is telling. The inspectors did not interview the chief electronics technician.[52]

The Chief Counsel's Report took a somewhat optimistic view of problems with the system, stating: "There is no evidence that the chairs malfunctioned on April 20."[53]

While that statement is technically correct, the chief electronics technician provided a more even-handed assessment during his deposition:

> *Q.* Is there—to your knowledge, based upon your familiarity with these systems, is there any way, sitting here today, to determine whether the A and B and C-Chairs were functioning correctly as of the time of the blowout on April 20th?
> *A.* No way to know.[54]

The Mud Logger's Office

During the production tail activity, the mud logger observed pit volumes, pressures, and flow readings from the Sperry acoustic flow sensor, and notified drillers of any indications of a kick. A screen gave the mud logger access to the camera in the shaker room, although he or she did not have access to any controls that could execute commands.

A single mud logger was on duty at all times. If that individual had to leave the office for any reason, he or she would notify crew in the drill shack by telephone. The split-tour system—in which only a subset of the crew changed at any one interval—helped ensure continuity of knowledge across tour changes. However, that system did not help if only one individual was on duty at a time, as was the case with the mud logger.

On most rigs, mud-logging companies install a second set of pit volume sensors to provide redundant readings of this critical measure. However, on the *Horizon,* mud loggers and drillers relied on readings from the same set of sensors. The mud logger testified that he had confidence in the information the system presented to him[55]—confidence that might not have been shared by a systems engineer who realized that the drillers and the mud logger relied on a single set of pit sensors, and that no one had calibrated them since the rig was built.

The Cement Unit

The cement unit had its own variable-speed, high-pressure pumps, pressure sensors, and gauges, and a small tank for measuring volumes of fluids bled off from the drill pipe. This equipment remained connected to the drill pipe by a *chiksan line*—a form of bendable high-pressure pipe—throughout the *Horizon*'s last hours. The senior WSLs typically monitored a negative test from this location—a practice that was not followed by the two WSLs who were on duty when Macondo blew out. Those WSLs relied instead on a member of the cementing crew to activate the cement pumps and report back the results of the first negative test by telephone.

The Shaker Room

The shaker room was directly below the drill shack, but there was no physical connection between the two. The room housed equipment in the flow path between the riser and the pits that was used to separate drill cuttings from mud during active drilling. The room had no displays or command capability, but it did house the HiTec flow meter, the Sperry flow meter, and the television camera. The crew in the shaker room could also directly observe flow out of the riser.

The equipment in the shaker room was normally tended by two crew members—who, because the equipment was so noisy, often resorted to hand signals to communicate. They used an intercom to communicate with the drill shack, and the noise level was so high that drillers often could not identify who in the shaker room they were talking to.

The piping in the shaker room was complex and understood by only a few crew members in the subsea department. This, and the fact that relevant documentation went down with the *Horizon,* means that the descriptions of the piping in various reports are inconsistent with each other and with the rig drawings in the Marshall Islands report.

What we do know is that the riser was capped with a device called a *diverter*—a rubber "donut" that, like the annular preventer, could close on a drill pipe or an open hole, but could resist much less pressure. Closing the diverter prevented returning fluid from rising up and onto the drill floor. Valves controlled either from the BOP panel or in the shaker room could be set to route that fluid to any of several destinations: directly overboard from one or both sides of the rig, through the mud-gas separator, or into the pits. Compounding the challenge of reconstructing the configuration of this equipment is the fact that the diverter on the *Horizon* was different from those on the rest of Transocean's fleet. And, like so much other equipment on the *Horizon,* it was in need of repair.[56]

Whatever the exact route of fluid returning from the well, the first piece of equipment it encountered was a device called the *gumbo box* (see Figure TN6.4). "Gumbo" is oilfield jargon for sticky shale that is hard to separate from drilling mud. During active drilling, the gumbo box is used to break such material up into manageable pieces. During the production tail activity, the box was basically a passive entity in the path between the riser and the pits. However, the gumbo box was significant because it provided a place where the flow stream came out in the open, it was where the Sperry Sun acoustic flow sensor was mounted, and it was the device used to shunt the flow overboard.

The upper portion of the diagram depicts the flow logic, and the lower is a notional representation of the physical box. The box was provided by NOV, the same vendor that provided the HiTec consoles. Transocean investigators did not know what model of gumbo box was on the rig, and they used

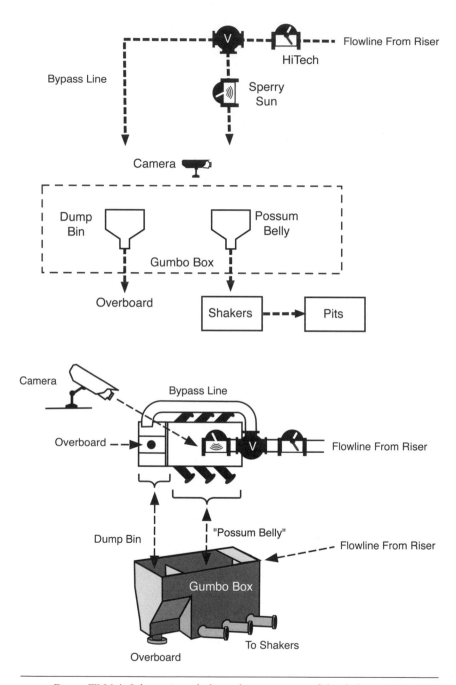

Figure TN6.4. Schematic and physical arrangement of the shaker room.

personal contacts within NOV to bypass that company's legal department to try to find out. Whether they ever did find out is unknown.[57]

The gumbo box had two compartments. One, called the "possum belly" because of its multiple outlets, contained the machinery to break up the gumbo. The processed material then went to the shakers, which separated it from the drilling mud, and from there into the pits. The second compartment, called the dump chute, had an outlet that led down through an opening in the bottom deck called the "moon pool."[58]

A valve at the entry to the box could be set to send fluid into the possum belly—and therefore to the pits—or through a bypass line into the dump box and overboard. The television camera was pointed at the outlet into the possum belly, as described by one of the senior WSLs when asked how he would react to an anomalous reading on pressures in the well:

A. I would go look in the gumbo box and see what the well was doing.
Q. And what's the gumbo box going to tell you?
A. Well, that's where . . . the flow comes in to the back of the gumbo box that sends it overboard. You can't see on the camera. The camera's facing in front of a gumbo box into the flow line. And when you're dumping overboard or switch to overboard mode to dump stuff, it goes into the back of the gumbo box and comes in low so you can't see it with the camera no more. You've got to go around to look—actually look in there.[59]

The arrangement of this camera and the location of the Sperry Sun flow sensor meant that the mud logger could believe that the well was quiescent when in fact it was flowing, unless notified by the drill crew as to how the piping was lined up.

Fluid such as mud and spacer leaving the well and the riser first encountered the HiTec paddle-style flow sensor and then the valve at the entry to the gumbo box, which determined further routing. The Sperry Sun sensor was installed on the other side of the valve. If the valve was set to send the flow to the pits, the lineup was as shown in Figure TN6.5.

In this configuration, the flow goes through both flow sensors and is visible to the camera. If the valve was set to route the fluid overboard, the signals from this portion of the circulating system could be misleading (see Figure TN6.6).

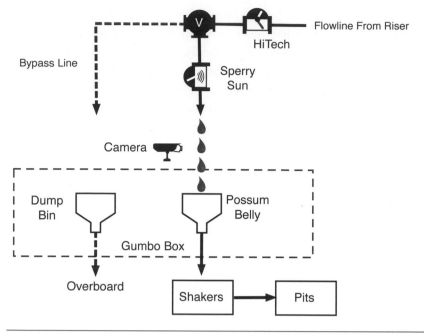

Figure TN6.5. The shaker room lined up as a "closed loop."

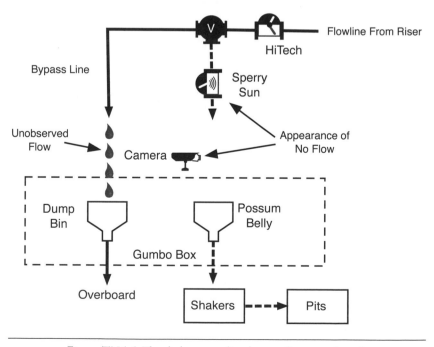

Figure TN6.6. The shaker room lined up to divert overboard.

In this configuration, the camera and Sperry Sun flow sensor could signal that there was no flow while in fact flow was occurring on a path they could not observe or sense. The HiTec sensor would be signaling the existence of the flow, but the mud logger did not have access to its output. The result was further disaggregation of information and potential for uncoordinated action.

Transocean investigators speculated that a workaround for this shortcoming was available:

> What is interesting to me is, if the Sperry sensor was located at the entry to the [possum belly], why would the crew want to use the bypass line to send the flows from the flow line directly overboard when they can just close the feeds to the shale shakers? The box is designed so that it will spill over into the [dump] chute if the volume entering the box exceeds the volume going out to the shale shakers causing the pool height to rise. If the shale shaker feed valves are closed, this would allow the distribution box to fill and spill over into the [dump] chute so that the Sperry sensor could still measure flow into the box.

The investigators attempted to verify this surmise, but there is no evidence of any resolution.[60]

Whether the workaround existed or not, it is difficult to believe that anyone would consciously include such a shortcoming in a designed control system. It is much more likely that convenience of installation had driven the ad hoc configuration of sensors and camera in the shaker room—a configuration that served to obscure rather than clarify what was going on down in the well.

Notes

I. INTRODUCTION

1. http://www.courthousenews.com/2010/07/22/29034.htm.

2. The Coast Guard awarded the Distinguished Public Service Award to the crew of the *Bankston,* a Certificate of Valor to her captain, and Silver Lifesaving Medals to the two crew members who staffed her recovery boat. The crew of the *Ramblin' Wreck* received a Public Service Award, and seven members of the *Horizon*'s crew received commendation for "selfless and heroic actions" during the rescue.

3. http://www.earthday.org/campaign/earth-day-2010-honoring-40-years.

4. U.S. Coast Guard Witness Statement dated 3:50 p.m., April 21, 2010.

5. MDL 2179 deposition taken March 14, 2011.

6. Hunter S. Thompson, *Hell's Angels: A Strange and Terrible Saga* (New York: Random House, 1966), 166.

7. The "brake" is a control used during drilling. We adopt the industry's informal term "driller" to mean the person in control of the equipment, whether the rig is actually "making hole" or doing other well construction tasks, which makes the observation equally valid for any time a rig is attached to a well.

8. The quote is from an anonymous Marathon Oil employee who prepared an internal briefing shortly after the release of BP's report of its internal investigation, in MDL 2179 Deposition Exhibit 7328.

9. Drillers must periodically attend well-control schools. These schools, typically a few weeks long at most, at the time were taught on the basis of lectures, workbooks, and multiple-choice tests.

10. Ernest K. Gann, *Fate Is the Hunter* (New York: Simon and Schuster, 1961), 109. Emphasis in the original.

11. MDL 2179 Deposition taken May 12, 2011. A rig is "latched up" to a well when it is connected to it by a riser.

12. The quotes are from BP's *"Deepwater Horizon* Accident Investigation Report, MDL 2179 Exhibit TREX-00001, and "Gulf of Mexico Oil Spill Response Fact Sheet" update of May 13, 2010 MDL 2179 Deposition Exhibit 1652.

13. Nancy Leveson, *Engineering a Safer World: Systems Thinking Applied to Safety* (Cambridge, MA: MIT Press, 2011), 17. Emphasis in the original.

14. Carroll, John S., "Incident Reviews in High-Hazard Industries: Sense Making and Learning under Ambiguity and Accountability," *Organization & Environment* 9 (June 1995): 175–197.

15. MDL 2179 Trial Exhibit TREX-07676; U.S. Department of the Interior, *Report Regarding the Causes of the April 20, 2010 Macondo Well Blowout,* Hereafter cited as *Department of the Interior Report,* http://docs.lib.noaa.gov/noaa_documents /DWH_IR/reports/dwhfinal.pdf; U.S. Coast Guard, *Report of Investigation into the Circumstances Surrounding the Explosion, Fire, and Sinking and Loss of Eleven Crew Members Aboard the Mobile Offshore Drilling Unit* Deepwater Horizon *in the Gulf of Mexico, April 20–22, 2010.*

16. Leveson, *Engineering a Safer World.* The paucity of technical evidence in the Macondo case precludes the full application of Leveson's analytical techniques.

17. The trial's formal name is 10-MD-2179 J, in re: Oil Spill by the Oil Rig Deepwater Horizon in the Gulf of Mexico on April 20, 2010.

18. See the Afterword for a complete discussion of sources and a description of important prior reports.

19. Vaughan, *The* Challenger *Launch Decision* (Chicago: University of Chicago Press, 1996).

20. Most notably, in the loss of the Space Shuttle *Challenger.*

21. Congressional Research Service, *Financial Performance of Major Oil Companies, 2007–2011,* February 17, 2012.

22. This analysis appears in the Afterword.

2. CONTROLLING MACONDO

1. MMS was renamed and reorganized in the immediate aftermath of the blowout. For simplicity we will, throughout this book, refer to the regulatory regime as MMS.

2. Private correspondence, confidential industry expert.

3. The name change to "BP" occurred in 2001.

4. John Podolny and John Roberts, *British Petroleum (A2): Organizing for Performance at BPX,* Case SIB-16A2, Graduate School of Business, Stanford University, April 2, 2002. Both the CEO in charge at the time of the blowout and his predecessor had close ties to the Stanford Graduate School of Business.

5. MDL 2179 Deposition taken March 23, 2011.

6. Quote from Tom Bergin, *Spills and Spin: The Inside Story of BP* (New York: Random House, 2011), 25.

7. Quote from John Roberts, *The Modern Firm* (Oxford: Oxford University Press, 2004), 190.

8. Bergin, *Spills and Spin,* 77. One of the early actions of the CEO who took over after the disaster was to accelerate the return to a more traditional model.

9. Ibid., 126.

10. Ibid., 77.

11. In his expert testimony, the same consultant quoted above characterized BP as a "brittle" corporation, one that was "at increased risk for major accidents when personnel are on vacation, sick or off work." MDL 2179 testimony given February 26, 2013.

12. http://www.csb.gov/u-s-chemical-safety-board-concludes-organizational-and -safety-deficiencies-at-all-levels-of-the-bp-corporation-caused-march-2005-texas -city-disaster-that-killed-15-injured-180/.

13. http://primis.phmsa.dot.gov/comm/reports/enforce/documents/520085044H /520085044H_Notice%20of%20Proposed%20Corrective%20Action%20Order 06202008.pdf.

14. http://www.publicintegrity.org/2010/05/17/2672/renegade-refiner-osha-says-bp -has-systemic-safety-problem.

15. Quote from Stanley Reed and Alison Fitzgerald, *In Too Deep: BP and the Drilling Race that Took it Down,* (New York: Wiley, 2011), 127.

16. MDL 2179 Deposition Exhibit 2250.

17. MDL 2179 Deposition taken May 19, 2011.

18. MDL 2179 deposition taken June 2, 2011.

19. A detailed discussion of this policy appears in Holley Doremus, "Through Another's Eyes: Getting the Benefit of Outside Perspectives," *Boston College Environmental Affairs Law Review* 38, no. 2 (2011): 247–281, http://lawdigitalcommons.bc .edu/ealr/vol38/iss2/3; and Oliver A. Houck, "Worst Case and the Deepwater Horizon Blowout: There Ought to Be a Law," *Tulane Environmental Law Journal* 24, no. 1 (2011), 1–18.

20. MDL 2179 Deposition Exhibit 2346. The response plan referenced in the exploration plan was that of November 2008. The exhibit is the revision of June 2009. The differences are probably not substantive.

21. Ibid., pages 11–7 and F-19. The same wording from the same contractor appeared in response plans accepted by MMS from other major operators in the Gulf, prompting awkward moments in a congressional hearing attended by their CEOs.

3. FROM RIG TO WELL

1. The two terms are synonymous in almost all contexts.

2. We explain the physics and computations of primary well control in Technical Note 1 in the Appendix.

3. MDL 2179 Deposition Exhibit 691.

4. MDL 2179 Deposition taken May 11, 2011.

5. More details on the *Horizon*'s BOP appear in Technical Note 2 in the Appendix.

6. http://online.wsj.com/articles/SB100014240527487033039045752932707464 96824.

7. Fred Bartlit Jr., Sambav N. Sankar, and Sean C. Grimsley, *Macondo: The Gulf Oil Disaster, Chief Counsel's Report* (National Commission on the BP Deepwater Horizon Oil Spill and Offshore Drilling, 2011), 237. Hereafter cited as *Chief Counsel's Report*. http://permanent.access.gpo.gov/gpo4390/C21462-407CCRforPrint0.pdf

8. These are also known as *annulars,* or *Hydrils,* after an early supplier.

9. Transocean, *Macondo Well Incident Transocean Investigation Report Volume II*, January 19, 2012, append. G, p. 168. https://web.archive.org/web/20120119045909 /http://www.deepwater.com/_filelib/FileCabinet/pdfs/12_TRANSOCEAN_Vol_2 .pdf

10. This situation has changed with the manufacture and prepositioning of subsea "capping stacks" by an industry consortium.

11. *U.S. Coast Guard Report.* Five more individuals were injured in the vicinity of the engines. Ten of the eleven fatalities occurred in close proximity to the blasts.

4. WEIGHING MACONDO'S RISKS

1. The lease number was G2306.

2. The technical measure is "boe" or "barrels of oil equivalent," which takes into account the possibility of natural gas.

3. MDL 2179 Deposition Exhibit 3203.

4. MDL 2179 Deposition of June 23, 2011.

5. *Chief Counsel's Report*, 25, and MDL 2179 Deposition Exhibit 6390.

6. A. Shadravan and M. Amani, "HPHT 101:What Every Engineer or Geoscientist Should Know about High Pressure High Temperature Wells," Society of Petroleum Engineers paper SPE 163376, December 2012.

7. MDL 2179 Trial Exhibits TREX-07500 and TREX-07501.

8. Compare, for instance, the fracture pressure estimates in MDL 2179 Deposition Exhibit 6290 with the analysis in MDL 2179 Trial Exhibit TREX-07510.

9. MDL 2179 Deposition Exhibit 6366.

10. MDL 2179 Deposition Exhibits 7059 and 7081.

11. *Chief Counsel's Report*, 268n7. The quote in the report is "a production well with an exploratory tail," which is clearly backward.

12. MDL 2179 Deposition taken June 28, 2011.

13. MDL 2179 Trial Exhibit TREX-41139.

14. MDL 2179 Trial testimony, February 26, 2013.

15. MDL 2179 Demonstrative D-4072.

16. For fuller discussion, see Leveson, *Engineering a Safer World,* 319–321.

17. See, e.g., *Chief Counsel's Report*, 271n84 and 297n159.

18. See, e.g., MDL 2179 Deposition Exhibit 0151, for reference to the "BTB coordinator," and 1125.

19. http://www.bloomberg.com/apps/news?pid=newsarchive&sid=a_9ZCw WCpY2k. The dollar figure is from http://cdn.exxonmobil.com/-/media/Reports /Corporate%20Citizenship%20Report/2010/community_ccr10_highlights.pdf.

20. MDL 2179 Exhibit TREX-45036.

21. See MDL 2179 Deposition taken April 22, 2011.

22. "Casing" is the high-strength steel tubing used to reinforce a well.

23. See, e.g., Leveson, *Engineering a Safer World,* pt. 3.

24. MDL 2179 Deposition Exhibit 6290.

25. MDL 2179 Trial Exhibit TREX-07502.

26. MDL 2179 Trial Exhibits TREX-07500 and TREX-07501.

27. MDL 2179 Deposition Exhibit 1552.

28. MDL 2179 Deposition Exhibit 3700.

29. Deposition of June 23, 2011. MDL Deposition Exhibit 3272 and Demonstrative D-4341.

30. MDL Deposition Exhibit 7083. "Pmean" is the average historical performance of nearby wells; "AFE" is the funds authorization, which contains the project cost estimate.

31. MDL 2179 Deposition Exhibit 3700.

32. MDL 2179 Deposition Exhibit 4066.

5. THE STRUGGLE WITH MACONDO

1. See Technical Note 3 in the Appendix for an explanation of how the MMS and BP measured drilling margin. For purposes of this chapter the reader may assume that the smaller the number, the less the margin.

2. MDL 2179 Trial Exhibits TREX-07510 and TREX-07511.

3. MDL 2179 Trial Exhibits TREX-08173 and TREX-07411.

4. MDL 2179 Deposition Exhibit 0691.

5. The detailed process is described in Technical Note 3 in the Appendix.

6. The rupture discs were a significant distraction to those attempting to stop the flow after the blowout when they did not know what path escaping oil was

taking—up the production casing or up the annulus between the production casing and the outside casings. If the latter, the discs could complicate the process of shutting off the flow of hydrocarbons from the well. Analysts later determined that the flow was moving up the production casing, and therefore that the discs did not present a problem.

7. As a rule, drilling engineers will try to avoid using a contingency casing unless absolutely necessary. Each casing they insert reduces the diameter of the completed well and increases the risk that the end of a practical "telescope" will be reached before the target depth.

8. MDL 2179 Deposition Exhibits 2376, 4751, and 1919.

9. National Commission on the BP *Deepwater Horizon* Oil Spill and Offshore Drilling, Transcript of Proceedings, November 8, 2010.

10. MDL 2179 Deposition Exhibit 7087.

11. MDL 2179 Exhibits TREX-07510 and TREX-07511.

12. MDL 2179 Deposition Exhibit 4757.

13. MDL 2179 Deposition Exhibit 1337.

14. MDL 2179 Deposition taken April 19, 2011.

15. National Hurricane Center, *Tropical Cyclone Report, Hurricane Ida* (AL112009), January 14, 2010.

16. JIT, August 23, 2010, and MDL 2179 Exhibit 1195.

17. John Konrad and Tom Shroder, *Fire on the Horizon: The Untold Story of the Gulf Oil Disaster* (New York: HarperCollins, 2011), 65. MDL 2179 Deposition Exhibits 1054, 1065 and 5545.

18. The involvement of the two other corporations was purely financial.

19. MDL 2179 Deposition Exhibit 1920.

20. MDL 2179 Deposition Exhibit 4008.

21. MDL 2179 Deposition of June 2, 2011.

22. MDL 2179 Deposition Exhibit 4500.

23. MDL 2179 Deposition of May 2, 2011.

24. These individuals were known for years in the industry as "company men," but the terminology used by BP is spreading.

25. See, e.g., MDL 2179 Deposition Exhibit 1127.

26. Ibid.

27. MDL 2179 Deposition Exhibit 0276.

28. Transcript, fifth meeting of the National Commission, November 9, 2010. Shell is known in the industry for heavy staffing levels. Private correspondence, confidential industry expert.

29. MDL 2179 Deposition taken May 15, 2011.

30. Norske Veritas, *Forensic Examination of Deepwater Horizon Blowout Preventer*, vol. 2 (Dublin, OH: Det Norske Veritas, 2011).

31. MDL 2179 Deposition Exhibits 1440, 1441, and 1442.

32. MDL 2179 Exhibit TREX-41139.

33. MDL 2179 Deposition Exhibit 1340.

34. MDL 2179 Deposition Exhibit 1552.

35. MDL 2179 Deposition taken May 3, 2011.

36. MDL 2179 Deposition Exhibit 1087. Total depth was the planned 19,650 feet.

37. The crew could mix the material in two stages, storing the first-stage mix until needed, and then adding the second-stage material, which thickens the pill, just before pumping it into the well. MDL 2179 Deposition Exhibit 2810.

38. MDL 2179 Deposition Exhibit 3065. Other MOCs had been processed, but they all documented personnel changes.

39. MDL 2179 Deposition Exhibit 1241.

40. MDL 2179 Trial Exhibits TREX-07510 and TREX-07511.

41. MDL 2179 Deposition Exhibit 6386.

42. There are minor inconsistencies between the last schematic submitted to MMS (MDL 2179 Deposition Exhibit 4754) and the BP's own report. Our diagram is from the latter.

43. MDL 2179 Deposition Exhibit 3540.

44. MDL 2179 Trial Exhibit TREX-05990, p. 24. The log is dated in the report as being from April 16, which is probably incorrect.

45. As a general rule, pore pressures increase with depth and the increasing weight of the rock above. Having a higher pore pressure at a shallower depth is rare but not unheard of.

46. MDL 2179 Exhibits TREX-08140 and TREX-60083.

47. MDL 2179 Deposition Exhibit 3538.

48. MDL 2179 Testimony given on March 4, 2013.

6. PLANNING THE PRODUCTION TAIL PHASE

1. This person was designated by court order as a Rule 30(b)(6) witness, which meant he answered for BP rather than as an individual. MDL 2179 Deposition Exhibit 5350.

2. MDL 2179 Deposition taken September 26, 2011.

3. http://www.nola.com/news/gulf-oil-spill/index.ssf/2010/07/27000_abandoned _oil_and_gas_we.html. There is essentially no difference in procedure between temporary and permanent abandonment.

4. Similar tests are called *negative pressure tests, negative tests,* or *inflow tests* across the industry.

5. MDL 2179 Deposition taken May 11, 2011.

6. MDL 2179 Deposition taken May 12, 2011.

7. MDL 2179 Deposition Exhibit 3466.

8. MDL 2179 Deposition taken May 11, 2011, lines 60:08 to 64:04.

9. *Chief Counsel's Report*, 301n40, 303n63, 304n87, 304n88.

10. The process of filling the drill pipe and kill line would pressurize the well. The first step was to bleed off this pressure, which would cause a small amount of fluid to flow.

11. MDL 2179 Deposition taken May 11, 2011.

12. Curiously, the "Rig Planner," a spreadsheet that combines a log of tasks performed with those planned, shows a mechanical plug called a *packer* to be installed rather than a cement plug. No explanation of this difference has been discovered. MDL 2179 Exhibit TREX-41139.

7. THE PRODUCTION CASING PLAN

1. MDL 2179 Deposition Exhibits 2561, 2575, and 2597.

2. MDL 2179 Deposition Exhibit 2458.

3. MDL 2179 Deposition Exhibit 2450. Emphasis in original.

4. Private correspondence, confidential industry expert.

5. The details of these and other mechanisms used in cementing are provided in Technical Note 4 in the Appendix.

6. In the oil industry, any length of casing or pipe shorter than normal is called a "sub."

7. MDL 2179 Deposition Exhibit 7087.

8. MDL 2179 Deposition of July 14, 2011.

9. MDL 2179 Depositions of June 16 and 17, 2011.

10. MDL 2179 Deposition taken June 1, 2011.

11. MDL 2179 Deposition Exhibits 1146, 2281, 2285, and 4223.

12. MDL 2179 Deposition Exhibit 2580.

13. MDL 2179 Deposition Exhibit 1687.

14. MDL 2179 Deposition Exhibit 1367.

15. In addition to the cited quotations, this section is based on MDL 2179 Deposition Exhibits 1517, 1684, and 1685 and JIT testimony given on October 7, 2010.

16. MDL 2179 Deposition Exhibit 1689.

8. THE CEMENTING PLAN

1. Bergin, *Spills and Spin*, 25.

2. The Chevron laboratory was used by the National Commission to duplicate Halliburton's testing of the cement used in Macondo. *Chief Counsel's Report*, Appendix D.

3. MDL 2179 Deposition Exhibit 2465. The assessment of OpiCem was not included in BP's *Report of Investigation* (MDL 2179 Exhibit TREX-00001).

4. MDL 2179 Deposition Exhibit 0626.

5. Ibid.

6. Ibid.

7. Exchange from MDL 2179 Deposition Exhibit 0136.

8. JIT, August 26 and October 7, 2010, and May 28, 2011; MDL 2179 Trial Testimony of April 2012, Depositions taken April 7 and 28, 2011, and Deposition Exhibits 007, 016, 224, 276, 300, and 358.

9. MDL 2179 Deposition Exhibit 1691.

10. ECD stands for "equivalent circulating density," the effective mud weight when the influence of fluid motion and other factors is taken into account. "Margin of safety between the ECD and the fracture pressure" is precisely equivalent to the phrase "drilling margin" used in Chapter 5.

11. National Research Council, *Macondo Well-Deepwater Horizon Blowout: Lessons for Offshore Drilling Safety* (Washington, DC: National Academies Press, 2011). http://www.nap.edu/catalog.php?record_id=13273.

12. MDL 2179 Deposition Exhibits 007, 153, 276, and 300.

13. Quotes from MDL 2179 Deposition Exhibit 1986. Emphasis in original.

14. "Plug and abandon," which means to set a bottom plug, as noted in National Research Council, *Macondo Well-Deepwater Horizon Blowout.*

15. MDL 2179 Deposition Exhibits 296, 319, 300, 4451. BP neither recorded nor obtained the services of a stenographer for its interviews, and the senior drilling engineer exercised his Fifth Amendment rights, so we have no further explication of these cryptic notes.

16. MDL 2179 Deposition Exhibit 0153.

17. MDL 2179 Deposition Exhibit 0151.

18. Neither BP's own report nor the Chief Counsel's Report mentioned the briefing slides or consideration of the "pure" abandonment option.

19. MDL 2179 Exhibit TREX-004456. Redrawn for clarity.

20. MDL 2179 Deposition Exhibits 0276 and 2035.

21. The nature of the configuration at the bottom of the well prevented the exercise of the other function of a CBL, which is to estimate the strength of the cement bond.

22. The ruling in MDL 2179 cited the fact that the BP plan relied on the inherently uncertain signals provided by the circulating system to indicate a proper cement job—and called for a CBL only if they did not—as an indication of reckless behavior on BP's part. MDL 2179 Ruling, para. 519.

23. MDL 2179 Trial Exhibit TREX-41139.

24. MDL 2179 Deposition Exhibit 1694.

9. THE TEST AND DISPLACEMENT PLAN

1. BP's subsea organization was the department responsible for undersea pipelines and other equipment that connected the wells, and should not be confused with the subsea specialists on the rig, who dealt with the riser and BOP.

2. MDL 2179 Deposition Exhibit 0835.

3. JIT, August 27, 2010.

4. In some parts of the world it is against regulations to abandon a well in an underbalanced state. Testimony before the National Oil Spill Commission, November 9, 2010.

5. It is now part of the regulations.

6. Private correspondence, confidential industry experts.

7. MDL 2179 Deposition Exhibit 0093. An "overpressured permeable section" is a porous formation where shove exceeds push—that is, where primary well control is not maintained.

8. MDL 2179 Deposition Exhibit 0674. Kill weight is the density of drilling fluid that guarantees that push is greater than shove—that is, that primary well control is maintained.

9. MDL 2179 Demonstrative D-8165.2.

10. As quoted in MDL 2179 Deposition Exhibit 5782.

11. MDL 2179 Deposition Exhibits 5782 and 0926.

12. MDL 2179 Trial Testimony, March 5, 2013. No justification was given for this additional delay.

13. Like most of the crew, the WSLs worked twelve-hour shifts. WSLs and some crew worked 6:00 a.m. to 6:00 p.m.; other crew members worked noon to midnight to provide overlap for continuous operations. Rig crews typically referred to a shift as a "tour"—often pronounced (and transcribed) as "tower."

14. MDL 2179 Deposition taken May 11, 2011.

15. MDL 2179 Deposition Exhibit 3568.

16. MDL 2179 Deposition taken May 11, 2011.

17. http://www.bp.com/en/global/corporate/press/speeches/subsea-containment -summit.html.

18. Private correspondence, confidential industry expert.

19. JIT, July 22, 2010, and Oct 7, 2010; MDL 2179 Trial Testimony, April 15, 2013.

20. MDL 2179 Trial Testimony, April 15, 2013.

21. Also known as the "PDQ" rig, for "production and drilling with crew quarters."

22. MDL 2179 Deposition Exhibits 0005 and 0063.

23. MDL 2179 Deposition Exhibit 3197.

24. Both the substitute and the junior WSLs were later indicted on federal man-slaughter charges. These charges were reduced to misdemeanor violations of the Clean Water Act. The junior WSL pleaded guilty and was sentenced to probation. The sub-stitute WSL chose to go to trial and was acquitted.

25. MDL 2179 Deposition Exhibit 4960.

26. MDL 2179 Deposition Exhibit 3555.

27. MDL 2179 Deposition Exhibit 0529. "T&A" is common jargon for "temporary abandonment," echoing "P&A" for "plug and abandon." "Logging operations" refers to the appraisal activities conducted after exploratory drilling ceased.

28. MDL 2179 Deposition Exhibits 836 and 1989. There are two versions of this document, entitled "Production Casing Operations," in the evidence. One is nineteen pages and the other twenty-one pages. Each appears as an attachment to two emails—identical down to the very second they were sent. Without access to the electronic versions it is impossible to determine how this curious situation of different attachments to two copies of the same email arose.

29. MDL 2179 Deposition Exhibit 0050.

30. MDL 2179 Deposition Exhibit 4242.

31. MDL 2179 Deposition Exhibit 4243.

32. MDL 2179 Deposition Exhibit 0537.

33. MDL 2179 Trial Exhibit TREX-01806.

34. *Chief Counsel's Report*, 338n78.

35. MDL 2179 Trial Testimony, March 5, 2013.

36. JIT, July 19, 2010.

37. MDL 2179 Deposition taken November 4, 2011.

38. JIT, July 19, 2010.

39. *Chief Counsel's Report*, 160; and *Department of the Interior Report*, 180.

40. Private correspondence, confidential industry expert.

41. The MDL 2179 ruling cited the decision to use an abnormal amount of LCM mixture as spacer as an instance of reckless behavior on the part of BP. MDL 2179 Ruling, para. 519.

42. MDL 2179 Deposition Exhibit 1019.

43. JIT, July 20, 2010.

44. MDL 2179 Deposition Exhibit 1134.

45. MDL 2179 Deposition Exhibit 4032.

46. MDL 2179 Deposition Exhibit 7392.

47. MDL 2179 Deposition Exhibit 0276.

48. MDL 2179 Deposition Exhibit 1396. This exhibit contains emails pertaining to another topic, and has an attachment placed out of sequence.

49. MDL 2179 Deposition Exhibit 1816.

50. Ibid. The format of time stamp on the email suggests that the well team leader was in his office on Sunday.

51. JIT, July 19, 2010.

52. The technology does not exist to directly measure flow at the pressures the mud pumps are capable of generating. Drillers therefore measure flow indirectly, by counting pump strokes and multiplying by the theoretical capacity of the pump cylinders, adjusted for mechanical and hydraulic inefficiency.

53. Videotaped testimony played in court on March 20, 2013.

54. Private correspondence, confidential industry expert.

55. MDL 2179 Deposition Exhibit 7607 and videotaped testimony played in court on March 20, 2013.

56. MDL 2179 Deposition Exhibit 4960.

57. JIT, May 26, 2010. The events of the 11:30 meetings of April 19 and 20 are conflated in the recollections of some of the witnesses who were able and willing to testify. Our reconstruction of the events is the one that in our judgment best fits the available evidence.

58. MDL 2179 Deposition Exhibit 0045.

59. MDL 2179 Deposition Exhibits 0319, 0041, and 4451.

60. JIT July 19, 2010.

61. MDL 2179 Deposition Exhibit 2821.

62. JIT, May 26, 2010; MDL 2179 Depositions taken January 29, May 3, July 20, and November 8, 2011, and Deposition Exhibit 5732. The OIM testified before the JIT that he did not recall making the statement.

63. MDL 2179 Deposition Exhibit 3575.

64. MDL 2179 Deposition Exhibit 3196.

10. A RECONSTRUCTED PLAN

1. We base our calculations on the assumption that the well was subject to a pore pressure of 11,936 psi, or, in the notation favored by drillers, 12.6 ppg equivalent mud weight (EMW) at 18,218 ft (Formation M56F).

2. MDL 2179 Exhibits TREX-05990 and TREX-08140.

3. MDL 2179 Trial Exhibit TREX-05990.

4. Ibid.

5. MDL 2179 Deposition Exhibit 0153. In addition to the cited works, this section was based on the Chief Counsel's Report Section 4.3 and notes; MDL 2179 Deposition Exhibits 0716A, 0717A, 0806, 2465, 2737, 5801, 5817, and 5824; and correspondence with confidential industry experts.

6. Private correspondence, confidential industry expert.

7. *Chief Counsel's Report*, 136–137.

11. THE SYSTEMS OF THE *HORIZON*

1. The Marshall Islands, a small archipelago in the Pacific, enjoys a special relationship to the United States owing to its having been used as an atomic test site in the 1950s.

2. Transocean never made up its mind what terminology to use for elements of the *Horizon*, e.g., often referring to the chief engineer as the maintenance supervisor. As a rule, crew members on the marine side insisted on being denoted by traditional maritime names.

3. MDL 2179 Deposition taken July 27, 2011.

4. MDL 2179 Deposition Exhibit 5618.

5. MDL 2179 Deposition Exhibit 4948.

6. MDL 2179 Trial Exhibit TREX-47361.

7. MDL 2179 Deposition taken July 20, 2011.

8. MDL 2179 Trial Testimony given March 14, 2013.

9. JIT, August 27, 2010; MDL 2179 Deposition taken April 26, 2011; and MDL 2179 Deposition Exhibit 0007.

10. Vaughan, *The Challenger Launch Decision*, 409–422.

11. The eleventh fatality was the starboard crane operator.

12. Transocean kept most of the up-to-date and detailed drawings of the *Horizon* on the vessel, and they were lost when it was, which leads to uncertainty about where certain items were located or their exact configuration.

13. The U.S. Coast Guard was not amused. See *U.S. Coast Guard Report*, 13, 53, and 135.

14. Ibid, xii.

15. MDL 2179 Deposition Exhibit 0597. The exhibit has been redacted, so the omission cannot be stated with certainty. However, given the importance of this topic, had it been covered, Transocean would likely have cited it in its defense.

16. In contrast to the more famous mayday call, a pan-pan call (pronounced "pon pon") indicates a state of "urgency" and is intended to notify potential rescuers that assistance may be needed soon.

12. UP TO THE EDGE

1. The exact mechanism by which the conversion signal is recognized and responded to by the float collar is described in Technical Note 4 in the Appendix. In less fragile formations, float collars are often converted by hand before the casing is run in and the resulting surge is tolerated.

2. MDL 2179 Deposition Exhibits 2559 and 2574. In oilfield jargon, records are called "tallies." A worker's personal diary of activity is called a "tally book."

3. JIT, July 19, 2010, and MDL 2179 Deposition Exhibit 1252.

4. MDL 2179 Deposition Exhibit 1252.

5. A Halliburton expert witness asserted that Sperry Sun data showed that the production casing had buckled under 140,000 pounds of compressive force but that the crew had not noticed. This assertion was the basis for the judge's decision to reduce Halliburton's liability to 3 percent. We have been unable to reconstruct the expert's reasoning. MDL 2179 Trial Testimony given April 4, 2013 and MDL 2179 Ruling, para. 152.

6. MDL 2179 Deposition Exhibits 0005, 0045, and 1252.

7. MDL 2179 Deposition Exhibits 0007 and 0041. Later Sperry Sun data show the standpipe pressure gauge values exactly duplicating those for the pressure gauge

on the cement unit when both were lined up to the drill pipe. These data cast doubt on the junior drilling engineer's assertion.

8. Cementer's testimony from JIT, August 24, 2010.

9. MDL 2179 Deposition Exhibit 2584.

10. MDL 2179 Deposition Exhibit 2117.

11. MDL 2179 Deposition Exhibit 0045.

12. MDL 2179 Deposition Exhibit 2117.

13. See, e.g., MDL 2179 Trial Exhibit TREX-05990 and *Chief Counsel's Report*, 98.

14. MDL 2179 Deposition Exhibit 0282.

13. THE FINAL DAY

1. Private correspondence, confidential industry expert.

2. An analysis of this evidence is given in Technical Note 5 in the Appendix.

3. MDL 2179 Trial testimony given March 5, 2013.

4. Ibid. A "task-specific THINK drill" was a detailed rehearsal of the steps to be performed.

5. MDL 2179 Deposition Exhibit 2807.

6. MDL 2179 Deposition Exhibit 0547.

7. MDL 2179 Deposition Exhibit 3203.

8. JIT May 10, 2010.

9. MDL 2179 Trial Exhibit 50150.

10. Private correspondence, confidential industry expert.

11. Ibid.

12. MDL 2179 Deposition Exhibit 1019.

13. One industry expert consulted by the authors feels this explanation is unlikely. (Private correspondence, confidential industry expert.) There are so many confounding factors, such as the condition of the 7-inch casing and the condition and placement of the cement, that a convincing explanation may never be forthcoming.

14. *Chief Counsel's Report*, 301n40, 303n63, 304n87, and 304n88.

15. MDL 2179 Deposition Exhibit 3651.

16. MDL 2179 Deposition Exhibits 3652 and 3683.

17. JIT July 19, 2010. Time from the log of the *Bankston*.

18. MDL 2179 Deposition Exhibit 0006.

19. MDL 2179 Deposition taken June 29, 2011.

20. JIT, May 28, 2010. Neither the senior toolpusher nor the OIM testified as to what they did by way of assistance.

21. *Chief Counsel's Report*, 01–302, notes 46 and 47; MDL Deposition Exhibit 3800.

22. MDL 2179 Deposition Exhibit 3188.

23. MDL 2179 Deposition Exhibit 3325.

24. MDL 2179 Deposition taken May 11, 2011.

25. *Chief Counsel's Report*, 302n47.

26. The trial judge for MDL 2179 ruled that in his opinion the "bladder effect" explanation originated with the substitute WSL and not from a Transocean employee. He further ruled that whatever the source of the explanation, it should not have been accepted. MDL 2179 Ruling, paras. 284, 287, 289, and 291.

27. Private correspondence, confidential industry expert.

28. MDL 2179 Deposition taken June 29, 2011

29. MDL 2179 Deposition Exhibit 3325.

30. JIT, May 28, 2010. The "surface plug" was the final cement plug to be set at 8367 feet.

31. It is not clear whether this meeting constituted a THINK drill as defined by Transocean; in any case, no survivor who was in the meeting testified that it did.

32. In addition to the previous citations, the reconstruction of events in this chapter was based on JIT testimony from May 28, July 19, August 24 and August 26, 2010; MDL 2179 Trial Testimony taken March 5, 2013; MDL 2179 Trial Exhibit TREX-07532; MDL 2179 Depositions taken June 20 and June 30, 2011; and MDL 2179 Deposition Exhibits 0005, 0012, 3574, 3575, 3650, 3807, 4365, 4963, 4964, and 6041.

33. JIT, July 20, 2010.

14. GOING OVER

1. MDL 2179 Deposition Exhibit 6044.

2. JIT, August 26, 2010.

3. MDL 2179 Deposition Exhibit 3575.

4. MDL 2179 Deposition Exhibit 0296. The existence of this telephone call was not mentioned during the JIT hearings, and BP chose not to mention it in the report they released in September 2010; but they did include evidence of it in the material they voluntarily provided to the chief counsel's team, and the call was described in the Chief Counsel's Report (176). The judge in MDL 2179 severely rebuked BP for the omission of the call from their report, calling that section "patently false" and BP's explanation at trial "untenable." MDL 2179 Ruling paras. 278–280.

5. MDL 2179 Deposition Exhibit 0006.

6. MDL 2179 Deposition taken May 12, 2011.

7. MDL Deposition Exhibit 0006.

8. The test, which involved placing a small amount of spacer in a bucket of water and noting that no oil slick, or "sheen," appeared, was not mandated by regulation but the junior WSL insisted it be done anyway.

9. JIT, May 28, 2010. No log of this phone call survived, but it is possible to deduce its time by comparing other testimony from the senior toolpusher with the Sperry data.

10. JIT, December 7, 2010.

11. Private correspondence, confidential industry expert.

12. Ibid.

13. MDL 2179 Trial Testimony given March 25, 2013.

14. JIT, May 28, 2010.

15. MDL 2179 Deposition Exhibit 4472.

16. JIT, May 28, 2010.

17. Besides the citations above, the events in this chapter were reconstructed from JIT, May 27, May 28, October 5, and December 7, 2010; MDL 2179 Trial testimony given March 5, 13, and 25, 2013; MDL 2179 Depositions taken March 14, March 15, and September 28, 2011; MDL 2179 Deposition Exhibits 0331, 2620, 3650, 4469, 4472, 4803, 4960, 5461, 5561, 5624, and 6044.

AFTERWORD

1. JIT, May 28, 2010.

2. MDL 2179 Deposition taken March 25, 2011, and MDL 2179 Deposition Exhibit 4365.

3. Marine Boards of Investigation are governed by 46 U.S. Code § 6301 and 46 C.F.R. § 4.01 et seq. See http://www.ecfr.gov/cgi-bin/text-idx?c=ecfr&rgn =div5&view=text&node=46:1.0.1.1.4&idno=46#46:1.0.1.1.4.6.

4. MDL 2179 Testimony for March 5, 2013.

5. Case 2:10-md-02179-CJB-SS Document 13355, Findings of Fact and Conclusions of Law, Phase One Trial.

6. MDL 2179 Trial Exhibit TREX-00559.

7. See http://www.bsee.gov/uploadedfiles/dnvreportvoli.pdf, http://www.bsee .gov/uploadedfiles/dnvreportvolumeii.pdf, and http://www.bsee.gov/uploadedFiles /DNVAddendumFinal.pdf.

8. See http://www.dsto.defence.gov.au/innovation/black-box-flight-recorder/david -warren-inventor-black-box-flight-recorder; http://www.heavytruckedr.org/diesel .html; http://papers.sae.org/2000–01–3551/; http://www.momentum-eng.com/MCLE -Automotive.html; and http://www.imo.org/ourwork/safety/navigation/pages/vdr.aspx.

9. See http://www.discovery-group.com/pdfs/HistoryWirelineCompanies.pdf.

10. Halliburton website, "INSITE Anywhere™—Watch Your Logs in Real Time—From Anywhere," May 7, 2008, archived at https://web.archive.org/web /20080507145131/http://www.halliburton.com/ps/default.aspx?pageid=454&navid =148&prodid=PRN::IYAG8YE8Z. The informal name for these data, and the retention of the Sperry Drilling name by Halliburton, has obscured the fact that one division of Halliburton provided evidence in a trial involving another.

11. See Technical Note 6 in the Appendix for the detailed configuration of the sensors on the *Horizon*.

12. The full charts may be accessed at http://www.uscg.mil/hq/cg5/cg545/dw /exhib/Sperry%20Data%20HAL_0048973%20.pdf.

13. "ASCII" generally refers to a scheme for coding letters and numbers in a computer. Sperry evidently co-opted the term for their digital, as opposed to analog, format.

14. Stress Engineering Services, "Hydraulic Analysis of Macondo #252 Well Prior to Incident of April 20, 2010," MDL 2179 Trial Exhibit TREX-50150.

15. NOV statement quoted in http://www.computerworlduk.com/news/it -business/3252262/bp-oil-spill-investigation-stranded-after-drilling-firm-refuses -software-access/ and http://fuelfix.com/blog/2010/12/06/presidents-spill-commission -complains-nov-wont-cooperate/.

16. Ibid.

17. As an indication of the lack of attention paid to issues surrounding the displays, the photograph on page 183 of the Chief Counsel's Report shows a model of HiTec display that is different from the model on the *Horizon*.

18. Available with supporting documents at http://www.uscg.mil/hq/cg5/cg545 /docs/documents/Deepwater.zip.

19. See http://docs.lib.noaa.gov/noaa_documents/DWH_IR/reports/dwhfinal.pdf.

20. MDL 2179 Deposition Exhibits 0001 and 0002.

21. Steve Coll, *Private Empire: ExxonMobil and American Power* (New York: Penguin Press, 2012), 28–34

22. E.g., "caused in part when a section of drill pipe buckled, which led to the malfunction of a supposedly fail-safe blowout preventer." http://www.nytimes.com /2015/04/11/us/new-sea-drilling-rule-planned-5-years-after-bp-oil-spill.html; http://www.nytimes.com/2015/04/14/us/new-regulation-aims-to-prevent-explosions -at-offshore-rigs.html; and http://www.nytimes.com/2015/05/12/us/white-house-gives -conditional-approval-for-shell-to-drill-in-arctic.html.

23. Available from http://cdm16064.contentdm.oclc.org/cdm/ref/collection /p266901coll4/id/3650.

24. http://www.gpo.gov/fdsys/pkg/GPO-OILCOMMISSION/pdf/GPO-OIL COMMISSION.pdf.

25. http://permanent.access.gpo.gov/gpo4390/C21462–407CCRforPrint0.pdf.

26. http://www.politico.com/news/stories/1110/44834.html.

27. *Chief Counsel's Report*, 246.

28. Ibid., 247.

29. MDL 2179 Trial Testimony given March 13, 2013.

30. MDL 2179 Trial Demonstrative D-8167.1.

31. MDL Deposition Exhibits 0093 and 1575.

32. Crews whose performance met or exceeded plan received bonuses of up to several thousand dollars.

33. MDL 2179 Deposition taken July 20, 2011.

34. MDL 2179 Deposition Exhibit 3188. In this and the quote above, the context makes it clear that "they" refers to BP Town.

35. *Chief Counsel's Report*, 248.

36. For example, the performance review for the well team leader mentioned "non-rig rate and logistics costs" and "reducing people and material costs." MDL Deposition Exhibit 7099.

37. http://www.washingtonpost.com/wp-dyn/content/article/2010/10/08/AR 2010100803145.html.

38. MDL 2179 Ruling, para. 72.

39. A Google search in July 2014 of "$58 million" *and* "Macondo" yielded more than 100,000 results.

40. All titles and organizations are as of the time of the loss. BP's management and organization have changed significantly since then.

41. MDL 2179 Deposition of June 27, 2011.

42. MDL 2179 Deposition Exhibit 2376.

43. MDL 2179 Deposition Exhibit 1919.

44. MDL 2179 Deposition Exhibit 6327.

45. MDL 2179 Deposition Exhibit 2370.

46. MDL 2179 Deposition Exhibit 2376.

47. MDL 2179 Deposition Exhibit 1920.

48. MDL 2179 Deposition Exhibit 2846.

49. MDL 2179 Deposition Exhibit 6331.

50. MDL 2179 Deposition Exhibit 1922.

51. BP denoted appraisal wells with the abbreviation IFT for "integrated flow test."

52. MDL 2179 Deposition taken June 28, 2011.

53. http://www.offshore-technology.com/projects/kaskida-field/.

54. These deadlines appear to be in place to prevent operators from stockpiling leases, thereby delaying the payment of royalties for oil extracted.

55. http://www.rigzone.com/news/article.asp?a_id=82447.

56. MDL 2179 Deposition Exhibits 6318 and 1488.

57. MDL 2179 Deposition Exhibit 1146.

58. MDL 2179 Deposition Exhibit 7087.

59. MDL 2179 Deposition taken July 14, 2011.

60. MDL 2179 Trial Testimony given April 15, 2013.

61. MDL 2179 Deposition Exhibit 1146.

62. MDL 2179 Deposition Exhibit 2285.

63. MDL 2179 Deposition taken July 15, 2011.

64. MDL 2179 Deposition taken May 19, 2011.

65. MDL 2179 Deposition Exhibit 2923.

66. *Chief Counsel's Report*, 355.

APPENDIX

1. Other organizations involved with drilling use specific gravity, or psi per foot. Ppg is the most common measure used in the Gulf of Mexico.

2. The number 0.052 is so basic that wearing it on an item of clothing will identify you as a member of the drilling community anywhere in the world.

3. The actual mud weight selected would depend on other factors, such as the pressure generated by the circulation of the mud.

4. The upper section is officially called the *lower marine riser package* or LMRP. The lower section, confusingly, is called the *BOP stack*. We adopt a simplified terminology for clarity.

5. MDL 2179 Trial Exhibits TREX-48102 and TREX-48103. These may be photographs of the BOP panel from the *Deepwater Nautilus* and not the *Horizon*.

6. MDL 2179 Deposition Exhibit 1868. A "bottom hole assembly" is typically the drill bit, which is much larger than the drill pipe upon which the preventer would be closed.

7. *Chief Counsel's Report*, 237.

8. MDL 2179 Deposition Exhibit 3600; and DNV report, vol. 2, p. F-11.

9. MDL 2179 Trial Exhibit TREX-04114.

10. The crew used the abbreviation as a verb, as in "to EDS."

11. *Systems Engineering Handbook*, INCOSE-TP-2003–016–02, version 2a, June 1, 2004, available at http://www.scribd.com/doc/49885991/Incose-SE-Handbook -2004#scribd.

12. MDL 2179 Trial Exhibit TREX-004271 and Deposition Exhibit 1488.

13. http://breakingenergy.com/2015/04/27/new-offshore-oil-regulations-respond-to -key-failures-of-deepwater-horizon-spill/.

14. MDL 2179 Trial Exhibit TREX-07536, append. H.

15. MDL 2179 Deposition taken May 15, 2011.

16. MDL 2179 Deposition taken June 30, 2011.

17. Roberts et al., Sutherland Asbill, & Brennan LLP, *Response to Coast Guard Draft Report by Transocean Offshore Deepwater Drilling Inc. and Transocean Holdings LLC, June 8, 2011.*Available at http://www.eenews.net/assets/2011/06/08/document _pm_01.pdf. The loophole has since been closed, and periodic teardowns and inspections are now mandatory.

18. In addition to the cited references, this note is based on MDL 2179 Trial Exhibits TREX-07535, TREX-07660, TREX-07687, TREX-07688, TREX-40008, TREX-40009, TREX-400020, TREX-61123, and TREX-61124.

19. MDL 2179 Deposition Exhibits 7510 and 7511.

20. MDL 2179 Deposition Exhibits 8173, 8174, and 7411.

21. Private correspondence, confidential industry expert.

22. Because the production casing used at Macondo had two diameters, Weatherford—the vendor of the plugs—had to have a custom set of top and bottom plugs made in an overseas factory.

23. MDL 2179 Deposition taken May 11, 2011.

24. MDL 2179 Trial Exhibit 50150.

25. MDL 2179 Trial Exhibits 07401, 40003, and Trial Testimony of April 9, 2013.

26. http://www.csb.gov/assets/1/19/Appendix_2_A__Deepwater_Horizon_Blow out_Preventer_Failure_Analysis1.pdf.

27. Phil Rae, *Deepwater Horizon Macondo Blowout,* InTuition Energy Associates Pte. Ltd., December 2010; and *The Genesis of the Deepwater Horizon Blowout Full Report,* http://www.energytribune.com/6337/the-genesis-of-the-deepwater-horizon -blowout-full-report#sthash.tLGYgqVg.dpbs. The original report is under revision and is no longer accessible.

28. *Chief Counsel's Report,* 301n42.

29. http://www.csb.gov/assets/1/19/Appendix_2_A__Deepwater_Horizon_Blow out_Preventer_Failure_Analysis1.pdf.

30. Private correspondence, confidential industry expert.

31. See Chapter 12.

32. JIT May 11, 2010.

33. The absence of such a cross-correlation of sensor readings in the Sperry equipment suggests that its designers lacked systems understanding.

34. MDL 2179 Demonstrative D-8249, and Trial Testimony given April 3, 2013.

35. MDL 2179 Deposition Exhibit 4820. The *Nautilus* was and is a near-duplicate of the *Horizon* that used anchors instead of dynamic position to maintain station.

36. Private correspondence, *Deepwater Horizon* crew member.

37. The judge in MDL 2179 ruled that by not reacting properly to these signals, "the Transocean drill crew failed to exercise proper well control." MDL 2179 Ruling, para. 336.

38. MDL 2179 Deposition Exhibit 4636.

39. MDL 2179 Deposition Exhibits 0658 and 4819.

40. JIT, July 20, 2010.

41. MDL 2179 Deposition Exhibit 4819.

42. MDL 2179 Deposition taken July 20, 2011.

43. Chief Counsel's Report, 308n56.

44. MDL 2179 Trial Testimony, March 5, 2013.

45. MDL 2179 Deposition taken July 20, 2011. The "tag replicator" is software, one of the functions of which is to restore sensor readings inside the computer after a crash.

46. MDL 2179 Trial Testimony given March 25, 2013.

47. MDL 2179 Deposition Exhibit 0275.

48. Ibid.

49. JIT, July 19, 2010, July 22, 2010, and December 9, 2010; and MDL 2179 Depositions taken August 17 and 18, 2011. JIT, August 23, 2010, and MDL 2179 Trial Testimony given March 5, 2013. MDL 2179 Deposition taken March 28, 2011.

50. MDL 2179 Deposition taken July 20, 2011.

51. MDL 2179 Deposition Exhibit 3285.

52. MDL 2179 Deposition taken July 20, 2011.

53. *Chief Counsel's Report*, 171.

54. MDL 2179 Deposition taken July 20, 2011.

55. JIT, December 7, 2010.

56. MDL 2179 Deposition Exhibits 5626 and 5627.

57. MDL 2179 Trial Exhibit 05125.

58. The "moon pool" was so named because someone standing on the bottom deck could see the reflection of the moon in it. Starting with the BP *Report of Investigation* (MDL 2170 Exhibit TREX-00001) most diagrams show the flow going overboard via a different route. Those diagrams cannot be reconciled with the rig drawings and survivor testimony.

59. MDL 2179 Deposition taken May 11, 2011.

60. MDL 2179 Deposition Exhibit 4934.

Acknowledgments

The authors are deeply indebted to Andrew Kay, who acted as our technical advisor and organized a group of reviewers that included David Hughes, Brian Wade, and Phil Rae of InTuition Energy Associates. We are similarly indebted to our editor Sandra Hackman, who was instrumental in converting the mass of material at our disposal into a coherent narrative. In addition, we draw on the insights offered within the writings of Robert Bea, Andrew Hopkins, and Charles Perrow. We would also like to thank M. de Lesseps for the perspective drawings in this book and Russell Boebert for the renderings on which they are based. As with any project of this magnitude, progress would have been impossible without the patience and support of our families.

Index